Developing Countries 2018

AA004719

Transportation Research Record: Journal of the Transportation Research Board

Volume 2672, Issue 31

ISBN: 978-1-5108-7834-1

Printed from e-media with permission by:

Curran Associates, Inc.
57 Morehouse Lane
Red Hook, NY 12571

Some format issues inherent in the e-media version may also appear in this print version.

Published by Sage Publications USA

ISBN (Print) 978-1-5108-7834-1
ISBN 2018 Printed Set (All Issues) 978-1-5108-7735-1

Printed by Curran Associates, Inc. (2019)

Additional copies of this publication are available from:

Curran Associates, Inc.
57 Morehouse Lane
Red Hook, NY 12571 USA
Phone: 845-758-0400
Fax: 845-758-2633
Email: curran@proceedings.com
Web: www.proceedings.com

Volume 2672 Number 31 December 2018

TRR
JOURNAL OF THE TRANSPORTATION RESEARCH BOARD

Contents

Developing Countries

Articles

Modeling the Gender Effects of Pedestrians and Calibration of the Modified Social Force Model 1
S. M. P. Siddharth and P. Vedagiri

Interactions between Driving Skills on Aggressive Driving: Study among Chinese Drivers 10
Chaozhong Wu, Wenhui Chu, Hui Zhang, and Türker Özkan

Cyclist Satisfaction Evaluation Model for Free-Floating Bike-Sharing System: A Case Study of Shanghai 21
Feifei Xin, Yifan Chen, Xiaobo Wang, and Xiaohong Chen

Study on the Invasion Behavior of E-Bikes with Motor Vehicle Traffic at a Signalized Intersection 33
Da Yang, Xiaoxia Zhou, Danhong Wu, and Sijing Liu

Influence of Socioeconomic Conditions on Crash Injury Severity for an Urban Area in a Developing Country 41
Rodrigo Mesa-Arango, Víctor G. Valencia-Alaix, Raul A. Pineda-Mendez, and Taleb Eissa

Macroscopic Spatial Analysis of Pedestrian Crashes in Medellin, Colombia 54
Cristhian Lizarazo and Víctor Valencia

Using Open Source Data to Measure Street Walkability and Bikeability in China: A Case of Four Cities 63
Peiqin Gu, Zhiyuan Han, Zhejing Cao, Yulin Chen, and Yang Jiang

Modeling Pedestrian Crossing Behavior and Safety at Signalized Intersections 76
S. Marisamynathan and P. Vedagiri

Performance Comparison of Bus Travel Time Prediction Models across Indian Cities 87
Jairam R, B. Anil Kumar, Shriniwas S. Arkatkar, and Lelitha Vanajakshi

How Operators' Legal Status Affects Safety of Intercity Buses in Thailand 99
Apiwat Ratanawaraha and Saksith Chalermpong

Measuring the Passenger Car Equivalent of Small Cars and SUVs on Rainy and Sunny Days 110
Majid Zahiri and Xiqun (Michael) Chen

Impact of Curbside Parking on Bicycle Lane Capacity in Nanjing, China 120
Xiaofei Ye, Xingchen Yan, Jun Chen, Tao Wang, and Zhao Yang

TRANSPORTATION RESEARCH BOARD 2018 EXECUTIVE COMMITTEE*

TRB

TRANSPORTATION RESEARCH BOARD

CORE SPONSORS

Transportation Departments of the 50 States and the District of Columbia

Federal Government

U.S. Department of Transportation
- Federal Aviation Administration (FAA)
- Federal Highway Administration
- Federal Motor Carrier Safety Administration (FMCSA)
- Federal Railroad Administration (FRA)
- Federal Transit Administration
- National Highway Traffic Safety Administration (NHTSA)
- Office of the Assistant Secretary of Transportation for Research and Technology (OST-R)

U.S. Air Force Civil Engineer Center (AFCEC)
U.S. Coast Guard (USCG)
U.S. Department of Energy
U.S. Department of the Interior (DOI)
U.S. Environmental Protection Agency (EPA)

Nongovernmental Organizations

American Public Transportation Association (APTA)
Association of American Railroads (AAR)
California Air Resources Board

The National Academies of
SCIENCES · ENGINEERING · MEDICINE

The **National Academy of Sciences** was established in 1863 by an Act of Congress, signed by President Lincoln, as a private, nongovernmental institution to advise the nation on issues related to science and technology. Members are elected by their peers for outstanding contributions to research. Dr. Marcia McNutt is president.

The **National Academy of Engineering** was established in 1964 under the charter of the National Academy of Sciences to bring the practices of engineering to advising the nation. Members are elected by their peers for extraordinary contributions to engineering. Dr. C. D. Mote, Jr., is president.

The **National Academy of Medicine** (formerly the Institute of Medicine) was established in 1970 under the charter of the National Academy of Sciences to advise the nation on medical and health issues. Members are elected by their peers for distinguished contributions to medicine and health. Dr. Victor J. Dzau is president.

The three Academies work together as the National Academies of Sciences, Engineering, and Medicine to provide independent, objective analysis and advice to the nation and conduct other activities to solve complex problems and inform public policy decisions. The Academies also encourage education and research, recognize outstanding contributions to knowledge, and increase public understanding in matters of science, engineering, and medicine.

Learn more about the National Academies of Sciences, Engineering, and Medicine at **www.national-academies.org.**

The **Transportation Research Board** is one of seven major programs of the National Academies of Sciences, Engineering, and Medicine. The mission of the Transportation Research Board is to increase the benefits that transportation contributes to society by providing leadership in transportation innovation and progress through research and information exchange, conducted within a setting that is objective, interdisciplinary, and multimodal. The Board's varied activities annually engage about 7,000 engineers, scientists, and other transportation researchers and practitioners from the public and private sectors and academia, all of whom contribute their expertise in the public interest. The program is supported by state transportation departments, federal agencies including the component administrations of the U.S. Department of Transportation, and other organizations and individuals interested in the development of transportation.

Learn more about the Transportation Research Board at **www.TRB.org**.

Article

TRR
JOURNAL OF THE TRANSPORTATION RESEARCH BOARD

Transportation Research Record
2018, Vol. 2672(31) 1–9

DOI: 10.1177/0361198118758673
journals.sagepub.com/home/trr
SAGE

Modeling the Gender Effects of Pedestrians and Calibration of the Modified Social Force Model

S. M. P. Siddharth[1] and P. Vedagiri[1]

Abstract

The design of pedestrian sidewalks depends on pedestrian flow, which is related to the speed of pedestrians on the sidewalk. The social force model (SFM) is a microscopic pedestrian simulation model that has been able to reproduce many self-organization phenomena of pedestrian flow such as lane formation. Studies have shown that the SFM has been modified to model particular pedestrian behaviors in different situations by introducing new forces or introducing new factors in existing forces. Also, the literature shows that pedestrian speed varies because of pedestrian characteristics such as age, gender, group behavior, and so forth. There are no studies that model the effect of these pedestrian characteristics using the SFM. Therefore, in this study, we have modeled the effect of gender of pedestrians by introducing a gender factor $\alpha_{m/f}$. A sidewalk in Mumbai, India has been chosen for this study. Pedestrian flow and speed were collected from the site. A base SFM containing the driving force, pedestrian–pedestrian interaction force, and pedestrian–boundary interaction force was coded in MATLAB. This model contains six parameters, which were calibrated using a genetic algorithm. Next, the SFM was modified to include different reaction times for the male and female pedestrians, τ_m and τ_f, respectively. Keeping other parameters as constant, τ_m and τ_f were calibrated and found. Gender factors α_m and α_f are found by dividing the reaction time τ_m and τ_f by τ_i, respectively. These gender factors could be found for the different male/female composition of pedestrians, which would help in analyzing the level of service of sidewalks.

Walking is one of the main traffic modes, especially for short trips. Walking is a major mode of travel especially in cities such as Mumbai in which 15 million trips are made on foot per day (*1*). A sidewalk or footpath is a facility catering to pedestrian movement adjacent to roadways (*2*). Pedestrian speed is an important parameter in analyzing these sidewalks as the width of the sidewalk is indirectly governed by pedestrian speeds (*3*). Pedestrian characteristics such as age, gender, and groups, and activities such as carrying baggage and using cell phones were found to cause variation in pedestrian speed. Several studies have found that the speed of male pedestrians is higher than that of female pedestrians (*4–9*). Middle-aged men were found to be faster compared with the elderly and young people (*5*). Pedestrian speed variation across age groups has also been observed in several studies (*6, 7*). The walking speed of people walking in groups was lower than that of people walking alone (*5*). Significant differences in speed between pedestrians carrying baggage and without baggage were also observed (*6*). Walking speeds in Western countries were found to be higher than in Asian cities because of cultural differences (*4*).

Modeling and simulating these pedestrian behaviors needs specialized models. Force-field analysis theory tells how a situation of a person is affected by others, especially in social situations (*10*). There can be two types of forces, helping forces and hindering forces, based on this theory. The idea of social force models (SFM) in modeling pedestrian behavior, using only a set of simple forces, is based on the concepts of this theory (*11*).

The SFM is a robust model and the variables are not arbitrary but rather have a physical meaning. The results of the model also show self-organizing phenomena. There are many modifications to the SFM that have been developed in various studies. Each study has taken a particular behavior of a pedestrian into account and modified the SFM to suit that behavior. Most of these studies were limited to evacuation situations or panic situations. There have been very limited studies for pedestrians in normal behavior using pedestrian facilities, considering pedestrian characteristics such as age, gender, and so forth. Therefore, the objective of the current

[1]Department of Civil Engineering, Indian Institute of Technology Bombay, Mumbai, India

Corresponding Author:
Address correspondence to S.M.P. Siddharth: siddharthsmp@gmail.com

study is to modify the existing SFM by introducing a factor to account for the gender of the pedestrian.

Literature Review

Base Social Force Model

SFM (*11–13*) assumes that the walking behavior of a pedestrian is influenced by a series of forces. According to the generalized force concept, the temporal change of location $x_i(t)$ of a pedestrian *i* obeys an equation of motion as shown in Equation 1:

$$\frac{dx_i(t)}{dt} = v_i(t) \tag{1}$$

where $x_i(t)$ is the position of pedestrian *i* at time *t* and $v_i(t)$ is the velocity of pedestrian *i* at time *t*. The velocity changes are as shown in Equation 2:

$$m_i \frac{dv_i}{dt} = F_i(t) + \xi_i(t) \tag{2}$$

where m_i is the mass of a pedestrian, $F_i(t)$ is the sum of forces influencing *i*, and $\xi_i(t)$ is the random fluctuations for behavioral variations at time *t*. The basic SFM for pedestrian simulation in normal conditions is given by Equation 3:

$$F_i = F_i^0 + \sum_j F_{ij} + \sum_B F_{iB} + \sum_\alpha F_{i\alpha} + F_g. \tag{3}$$

These forces are the desired movement force F_i^0, the repulsive social interaction force between pedestrians F_{ij}, the attractive social force toward another object $F_{i\alpha}$, and the repulsive social force from the walls F_{iB}. The model also contains an attractive social force F_g, allowing, for example, a group of pedestrians to stick together when moving around.

The desired movement force represents the pedestrians desire to move toward its target. A pedestrian, if unhindered by other means, will move toward its target with an initial desired speed that is given as a model parameter. The desired movement force is given by Equation 4:

$$F_i^0 = \frac{\left(v_i^0 e_i^0 - v_i\right)}{\tau_i} \tag{4}$$

where v_i^0 is the desired force, which is a parameter of the model, e_i^0 is the desired direction, v_i is the current speed of pedestrian at time *t*, and τ_i is the relaxation time, which is also a parameter of the model. This force models the desire of the pedestrians to adapt the current velocity to the desired velocity within a certain relaxation time. The relaxation time shows how fast a pedestrian can adapt current velocity to desired velocity.

The repulsive force between pedestrians is the force between all pedestrians in the model. The repulsive force of pedestrian *j* on pedestrian *i* is given by Equation 5:

$$F_{ij} = V * w(\varphi_{ij}) \tag{5}$$

where V is the force term, which can have various specifications. The first simple specification is given by Equation 6:

$$V\left(d_{ij}\right) = A_i e^{\left[\frac{\left(r_{ij} - d_{ij}\right)}{B_i}\right]} n_{ij} \tag{6}$$

where $d_{ij} = \left|x_i(t) - x_j(t)\right|$ is the distance between two pedestrians *i* and *j*, $r_{ij} = r_i + r_j$ is the sum of radii of the pedestrians *i* and *j*, and $n_{ij} = \left(x_i(t) - x_j(t)\right) / d_{ij}$ is the vector pointing from pedestrian *j* to *i*. A_i is the interaction strength and B_i is the range of repulsive interactions, which are parameters of the model. The second specification of the repulsive force is given by Equation 7, *a* and *b*:

$$V(b) = A_i e^{\left[\frac{(b)}{B_i}\right]} n_{ij} \tag{7a}$$

$$2b = \sqrt{\left(d_{ij} + d_{ij} - v_j \Delta t e_j\right)^2 - (v_j \Delta t)^2} \tag{7b}$$

This repulsive force is an elliptical force with semi-minor axis *b*. $v_j\,\Delta t$ is the stride of pedestrian *j*. The factor $w(\phi_{ij})$ is an angular modification that changes the way that force affects the pedestrians at the front and back of a group, which is given by Equation 8:

$$w(\phi_{ij}) = \left(\lambda_i + \left(1 - \lambda_i\right) \frac{\left(1 + \cos\phi_{ij}\right)}{2}\right) \tag{8}$$

where ϕ_{ij} is the angle between the desired direction e_i and the direction between pedestrian *i* and *j*, n_{ij}. It can be found using the relation $\cos\phi_{ij} = -n_{ij}.e_i$. $0 \leq \lambda \leq 1$ is a parameter of the model. When $\lambda = 1$ no angular modification is made. When $\lambda < 1$ the force is modified by a factor determined by the angle. This factor gives higher weightage to angles closer to 0 and lower weightage to angles closer to π. The attractive force $F_{i\alpha}$ has the same formulations as the repulsive force between pedestrians as shown in Equation 9:

$$F_{i\alpha} = A_{i\alpha} e^{\left[\left(r_{i-d_{i\alpha}}\right)/B_{i\alpha}\right]} n_{i\alpha} * \mathrm{w}\left(\varphi_{i-}\right) n_{i\alpha}. \tag{9}$$

However, their interaction range $B_{i\alpha}$ is negative and larger, and the strength parameter $A_{i\alpha}$ is smaller and time dependent, that is, it decreases with time. The repulsive force between pedestrians and boundaries or obstacles is given by Equation 10:

$$\vec{F}_{iW} = A_{iW} e^{\left[(r_i - d_{iW})/B_{iW}\right]} n_{iW} \quad (10)$$

where r_i is the radius of pedestrian i and d_{iW} is the distance between the nearest edge of boundary W for pedestrian i. Similar to the repulsive force between pedestrians, A_{iW} is the interaction strength and B_{iW} is the range of repulsive interaction. Finally, the attractive force between groups is given by Equation 11, allowing, for example, a group of pedestrians to stick together when moving around:

$$F_g = -C_{ij} n_{ij}. \quad (11)$$

The basic SFM was modified in different studies to increase the efficiency of modeling as well as to model different pedestrian behaviors.

Modifications to the Base Social Force Model

Several research studies had been carried out by modifying the base SFM model, especially for evacuation scenarios and panic conditions. Researchers had proposed an overlap eliminating algorithm (OLE) along with three modifications to the basic SFM. Also, a new parameter called the excitement parameter E was introduced to modify the current velocity in panic conditions. The three new modifications suggested in this model were density dependence of the social force, oriental dependence, and memory effects. This model is referred to as the LKP model based on the authors Lakoba, Kaup, and Finkelstein (*14*). In further studies, this LKP was simulated for pedestrians moving in a walkway and it was found that pedestrians were short-sighted; that is, they were not able predict crowds or obstacles as a normal pedestrian would do. Therefore, the authors introduced an investigative capability for pedestrians, who would look ahead and scan for areas of high densities and low densities so as to avoid crowds. Another study introduced a prediction factor to the previous LKP model. Pedestrians who have this prediction capability will launch their investigation before coming into contact with obstacles or crowded situations (*15*). Researchers then modified the LKP model to introduce a familiarity factor. This differentiates between pedestrians who are familiar with the exits and who are not familiar (*16*). Another major modification carried out in LKP model was assuming an elliptical shape for pedestrians for a single pedestrian row. In this the distance between pedestrians and pedestrians with boundaries will vary depending on the direction because of elliptical assumptions. The model was able to reproduce a fundamental diagram of the experimental data (*17*).

Studies also investigated the effect of the number and positions of evacuation leaders on the evacuation dynamics in rooms with limited visibility range. In this improved SFM, only the trained leaders exactly know the exit positions, and the others could only follow the guidance according to the positions and directions of the evacuation leaders (*18*). Further, researchers found that the formulation of normal behavior of pedestrians for evacuation conditions lacked a self-slowing mechanism for pedestrians. Therefore a respect distance was assumed around a pedestrian: if a pedestrian j enters the respect area of another pedestrian i then the desired velocity of i becomes zero. Only when the respect area is free does the desired velocity go back to its original value. It should be noted that the current velocity is not zero and there is no sudden stop movement for the pedestrian (*19*). A modified SFM for evacuation conditions that models stop areas and also includes a factor for psychological conditions was also proposed (*20*).

In this particular study, extended decision field theory was used on a tactical level and the SFM was used on a behavioral level. The base SFM considers the interactions between pedestrians and with boundaries known as psychological forces and also the physical forces. In addition, the grouping behavior of pedestrians is modeled. The study also considers the vision level of pedestrians. To model the vision range, a factor called the connection range (CR) is introduced. A pedestrian is affected only by other pedestrians within this CR. This study, which is for a shopping mall, has concluded that the modifications have helped the base SFM to create more realistic simulations (*21*).

Research studies also had investigated a one-dimensional SFM, that is, for a pedestrian moving in a single row, under experimental conditions. Previous studies have shown that for one-dimensional pedestrian flow the required length d for a pedestrian to move with velocity v is linearly related to the velocity itself. It has been found that this model was able to produce a fundamental diagram similar to one modeled from field data (*22*).

Further research studies observed that the original SFM has unrealistic simulations. The author had observed that when pedestrians move toward a wall, a pedestrian with a higher velocity experiences a higher collision force compared with a pedestrian with lower speed. In reality, a pedestrian will be able to predict the collision and will be able to slow down. Therefore a relative velocity term is introduced into the physical force term. Evacuation of pedestrians was simulated using this model and was found to be realistic (*23*). In another study, the interactions between pedestrians and attractive areas such as shops, street performance, and so forth, were modeled. If the attractive forces are strong enough pedestrians may stop moving to their destination and group around these attractions. This attraction force will have a repulsion component so that pedestrians do not go too near to the attractions. An imaginary corridor of length 25m was simulated with attractive interactions, equally spaced from each other and also for varying densities. This study was able to produce the expected realistic simulations (*24*).

In further research, it was found that a relation between the pedestrians' perception and the local density exists, which could be modeled with the help of fundamental diagrams (*25*). The SFM was further modified to model a pedestrian moving on stairs. In addition to the existing forces, a new gravity force is added to it based on basic mechanics. Also a factor is multiplied to the desired force to account for the change of angle of elevation (*26*).

Few studies had modeled the grouping behavior of pedestrians and included in the SFM. The authors considered the pedestrians moving in groups of 2, 3, and 4 and replicated their orientation and their distance between each other. Three force terms are introduced in this model. The first is to model the pedestrians turning their gazing angle toward the members of the group. The second is the attraction force toward the center of mass of the group and the third a repulsive force to prevent overlapping of pedestrians. This model was simulated, and the orientation angle and distance between pedestrians during simulations were compared and were found to be statistically similar (*27*).

Calibration of the Social Force Model

Researchers calibrated the SFM has been macroscopically with fundamental diagrams for pedestrians moving in experimental conditions (*17, 22*). In further studies, evolutionary algorithms were also used to calibrate the SFMs. Here trajectories were extracted from videos with the help of extracting software. Videos from controlled experiments were used. After extraction a series of simulations were carried out in which one chosen pedestrian was allowed to be simulated based on the SFM, whereas the remaining pedestrians moved exactly based on their trajectories. The error between the simulated trajectory and the actual trajectory of the single pedestrians is the fitness function. This is repeated for different pedestrians and for different time periods by varying the parameters. The best parameter set is found with the help of a genetic algorithm (*28*).

In another study, pedestrian trajectories were extracted by mouse clicking the pedestrian movement on a frame of the video. Data from a transfer hub in Beijing was used. The coordinates were then transferred to real-world coordinates. v_i^0 and τ_i can be calculated directly by extracting pedestrian speed data from the video. To find the values of A_i and B_i, trajectories of a pair of pedestrians who were not influenced by boundaries or obstacles were chosen. The SFM was transformed in such a way that these values can be extracted based on a least squares method (*29*).

Researchers had estimated the social force parameters based on a data assimilation method (*30*). Another study had used the maximum likelihood estimation method to estimate the parameters of the SFM for pedestrians crossing a crosswalk. Here, 53 pedestrians for low density conditions and 105 pedestrians for high density conditions were taken (*31*). The same method was used to estimate the SFM parameters with the help of maximum likelihood estimation for a crosswalk in Japan (*32*).

Existing studies have modified SFM by introducing new forces or factors to model a particular observed phenomenon in their study location. Most of these studies have been in evacuation situations or panic conditions. Data in most of the studies was collected under experimental conditions, even in studies carried out for normal conditions. There are no studies that model the pedestrian characteristics of age and gender. Therefore, in this study we modify the base SFM by introducing a factor to take into account the gender of the pedestrian.

Modified Social Force Model Considering Gender

For our study, from Equation 3 we consider only three forces in the base SFM consisting of the driving force, the force between pedestrians, and the force between pedestrian and boundary:

$$F_{i(t)} = F_i^0 + \sum_j F_{ij} + \sum_B F_{iB}. \tag{12}$$

For the modified SFM we have the same three forces but the reaction time τ_i in the driving force is replaced by τ_m for male pedestrians and τ_f for female pedestrians as shown in Equation 13:

$$F_i^0 = \frac{\left(v_i^0(t)e_i^0(t) - v_i(t)\right)}{\tau_{m/f}}. \tag{13}$$

The gender factor $\alpha_{m/f}$ is calculated by dividing $\tau_{m/f}$ by the reaction time for all pedestrians without considering gender, that is, τ_i, as shown in Equation 14:

$$\alpha_{m/f} = \tau_{m/f} / \tau_i. \tag{14}$$

To find the gender factor the reaction time τ_i is found by calibrating the base social force considering all pedestrians as a single category. In the second step, $\tau_{m/f}$ is found by assuming different reaction times for males and females. From these reaction times the gender factor is calculated.

Data Collection and Speed Analysis

Field Data Collection

To model a modified SFM for sidewalks, a sidewalk in Mumbai, India was chosen. Pedestrian data were collected from a sidewalk in Mahapalika marg near Chathrapathi Shivaji Terminus (CST) in Mumbai. The site is located around 700 m from CST station. Geometric data were measured with the help of a roadometer. The width of the sidewalk is around 7 m. A section, of around 18.6 m in between two bus stops, was chosen for study from the site as shown in Figure 1. This trap length was marked on the facility with the help of self-adhesive tape. The flow of pedestrians in this section was bi-directional from entry 1 to 2 and 2 to 1 but a small number of pedestrians also entered and exited the site through entry 3 and 4 as shown in Figure 1.

A videography technique was used to record the movement of pedestrians. Videos were collected from the site during peak hours. Two video cameras, C1 and C2, were fixed

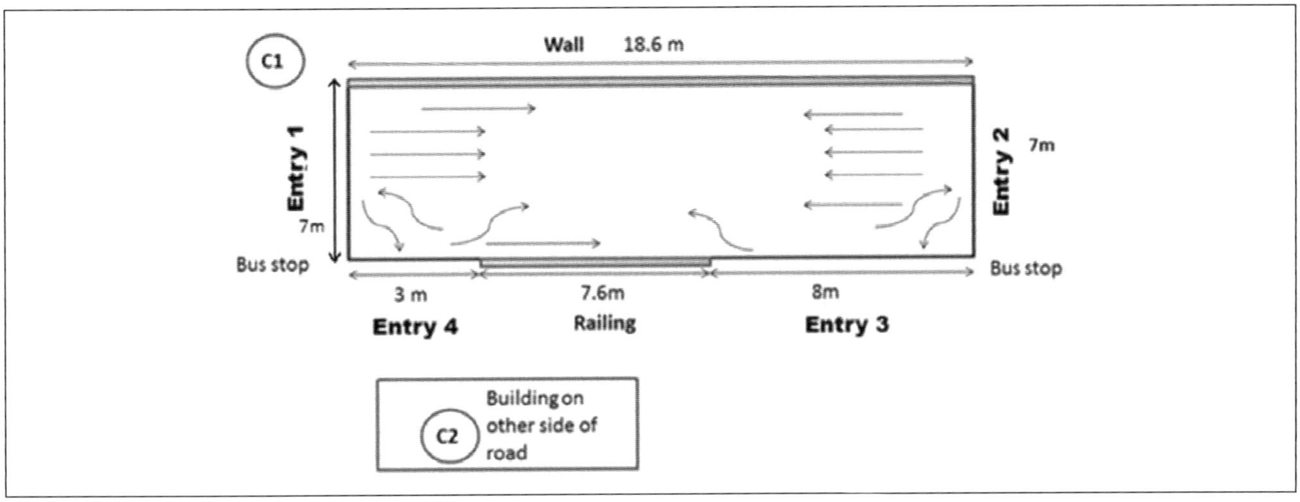

Figure 1. Geometry of the sidewalk.

Table 1. Pedestrian Speed Variation-Gender Wise

	Direction 1–2		Direction 2–1		Bi-directional	
Characteristic	No. of pedestrians	Avg. speed m/s	No. of pedestrians	Avg. speed m/s	No. of pedestrians	Avg. speed m/s
Male	312	1.26	219	1.22	531	1.24
Female	78	1.12	69	1.03	147	1.08

to collect data as shown in Figure 1. Camera C1 was fixed on a wall near entry 1 covering the pedestrians entering and exiting 1, and camera C2 was fixed on the roof of a building on the opposite of the road covering the whole section. From camera C1 we were able to identify the pedestrian movement in entry 1 and 4 and also the pedestrians' visual appearances (age, gender), type of luggage, cell phone usage, and group size. From camera C2 we were able to identify pedestrians in 2 and 4 and also extract the time taken by pedestrians to travel from 1 to 2, 1 to 3, 1 to 4, 2 to 1, 2 to 3, 2 to 4, 3 to1, 3 to 2, 4 to 1,4 to 2, and 4 to 3. The pedestrian data in the first half an hour was extracted. In total 800 pedestrians entered from all the sections (1, 2, 3, and 4) together. Of these, 431 pedestrians entered from 1, 318 pedestrians from two 2, 12 pedestrians from 3, and 39 pedestrians from 4. For all these pedestrians their characteristics (gender, age, luggage), and activity (cell phone usage and group size) were noted. Also the entry time at the start of the entrance and the exit time at the point pedestrians leave the sidewalk were also noted.

Pedestrian Speed Analysis

From the entry and exit time and length of the trap, the speed of individual pedestrians was calculated. The average speed of bi-directional flow was found to be 1.20 m/s. The average speed of direction 1–2 (1.23 m/s) was higher than the average speed of direction 2–1 (1.17 m/s). In all three cases, that is, direction 1–2, direction 2–1, and combined bi-directional flow, the following common trends were observed. In this study, only speed according to gender variations was considered. Male pedestrians were found to have a very high average speed compared to female pedestrians as shown in Table 1. An analysis of variance (ANOVA) test was conducted to find out if the bi-directional pedestrian speed is affected by gender. Statistical Package for the Social Sciences (SPSS) software was used to conduct ANOVA testing. These tests were carried out using a 95% confidence interval. Gender was the factor that was tested. The F calculated was 80.61, which was higher than the F critical value of 3.84 at a p-value of .00. Therefore, gender was found to significantly affect the speed of pedestrians.

Calibration of the Social Force Model

Flow per minute was extracted from the video data in both directions. The inflow from entry 1 and 2 were fitted with various distributions and the best distribution was found. Then the basic SFM as shown in Equation 12 was coded in MATLAB without considering the values of $\alpha_{m/f}$, that is, in the first step gender was not distinguished. The model was calibrated using a genetic algorithm and values of the six parameters of the model were obtained. In the second step, keeping the values of all the remaining five parameters, $\tau_{m/f}$ was calibrated. From this constant, $\alpha_{m/f}$ was calculated.

Table 2. Distribution Ranking for Direction 1–2/Direction 2–1

Serial No.	Distribution	Direction 1–2 Kolmogorov–Smirnov Rank	Direction 1–2 Anderson–Darling Rank	Direction 2–1 Kolmogorov–Smirnov Rank	Direction 2–1 Anderson–Darling Rank
1	Discrete uniform	3	4	3	4
2	Geometric	4	3	4	3
3	Logarithmic	5	5	5	5
4	Negative binomial	1	1	2	1
5	Poisson	2	2	1	2

Table 3. Negative Binomial Statistics for Both Directions

Direction 1–2						Direction 2–1					
Kolmogorov–Smirnov Test											
Sample size			30			Sample size				30	
Statistic			0.10			Statistic				0.15	
p-value			.83			p-Value				.39	
Rank			1			Rank				2	
α	0.2	0.1	0.05	0.02	0.01	α	0.2	0.1	0.05	0.02	0.01
Critical value	0.19	0.21	0.24	0.27	0.28	Critical value	0.19	0.21	0.24	0.27	0.28
Reject?	No	No	No	No	No	Reject?	No	No	No	No	No
Anderson–Darling Test											
Sample size			30			Sample size			30		
Statistic			0.55			Statistic			0.75		
Rank			1			Rank			1		
A	0.2	0.1	0.05	0.02	0.01	A	0.2	0.1	0.05	0.02	0.01
Critical value	1.37	1.92	2.50	3.28	3.90	Critical value	1.37	1.92	2.50	3.28	3.90
Reject?	No	No	No	No	No	Reject?	No	No	No	No	No

Model Inputs

The sidewalk considered was the study site near CST station, which contains two permeable entry/exits in addition to entry 1 and entry 2. The width of the footpath is 7 m and the length is 18.6. In this simulation, the permeable boundary was neglected and only pedestrians entering/exiting from entry 1/entry 2 were considered. A simulation time of 15 min was taken. 175 pedestrians moved from entry 1 to 2 and 144 pedestrians moved from entry 2 to entry 1. For both the entries, origin/destination area was created. These origin/destination areas were created 5 m from both the entries. In these areas the width of entry point is divided into equal parts and pedestrians were input into the footpath from these points. Thus the x-coordinate and y-coordinate of the entry of pedestrians were fixed. For each pedestrian, one of these x- or y-coordinates was chosen randomly. This was done for both the entries. The next input value to be entered was the initial velocity in the x and y directions. These values were randomly generated depending on the average speed and lower speed limit of pedestrians.

Inflow Distribution Fitting

Five discrete distributions were fit for the flow with the help of Easyfit software (*33*) and they were ranked for direction 1–2 and direction 2–1 as shown in Table 2. Kolmogorov–Smirnov and Anderson–Darling tests were conducted to find the best fit. For direction 1–2, both tests found that a negative binomial test gave the best fit as shown in Table 2. Table 3 gives the statistics for the Kolmogorov–Smirnov test as 0.10. This statistics value is less than the critical values for all the four values of α. For the Anderson–Darling test the statistic is 0.55, which is less than the critical values for all values of α. Therefore, the null hypothesis is accepted that the negative binomial is a good fit of the headway.

For direction 2–1 the best fit as per the Kolmogorov–Smirnov test was the Poisson distribution and the negative binomial was second best, whereas the Anderson–Darling test gave the negative binomial as the best fit and the Poisson as second best, as shown in Table 2. As the best fit in direction 1–2 was the negative binomial the same was also taken for direction 2–1. Table 3 gives the statistics for the

Table 4. Range of Parameters and Calibrated Values of the Basic SFM

Parameter	Range of parameters	Calibrated values
Relaxation time τ_i	0.1–1	1
Pedestrian–pedestrian interaction strength A_i	0.1–10	5
Pedestrian–pedestrian interaction range B_i	0.1–1	1
Boundary interaction strength A_{iW}	0.1–10	0.17
Boundary interaction range B_{iW}	0.1–1	7.27
Λ Anisotropy parameter	0.1–1	0.79

Kolmogorov–Smirnov test as 0.15. This statistics value is less than the critical values for all the values of α. For the Anderson–Darling test the statistic is 0.75, which is less than the critical values for all values of α. Therefore, the null hypothesis is accepted that the negative binomial is a good fit of the headway.

Calibration of the Base SFM

The range of parameters of the basic SFM is as shown in Table 4. The desired speed was kept as 1.34 m/s. In this table, based on the previous studies, the ranges of the parameters are fixed. MATLAB has an optimization toolbox called Optimtool. A genetic algorithm is available as one of the optimization tools. As manual calibration may not be able to find the least error, we used a genetic algorithm. Here the inputs required in the genetic algorithm interface are the fitness function, No of variables, bounds for these variables, and stall limits. The absolute percentage error between the simulated speed and the field speed was taken as the fitness function. As the number of variables is six the population in each generation was kept as 200, which was the default if the number of variables is more than five. The simulation was run for 15 min. The error was found every 5 min and then the average found. The best parameter values generated in the genetic algorithm are as shown in Table 4.

The calibration error was found to be 6.33%. For validation the simulation was run for the next 15 min flow with the same calibrated parameter set. The error was found to be 9%. Therefore, the model was validated.

Calibration of the Modified SFM

The modified social forces model has $\tau_{m/f}$ as the new parameter instead of τ_i. For all the male pedestrians τ_m is the relaxation time parameter whereas τ_f is the relaxation time for female pedestrians. Keeping all parameters found in the previous calibration as constant, the calibration was run with the same input values. The range of $\tau_{m/f}$ was kept as 0.1–1. Absolute percentage errors between the simulated and field speed of males and females were calculated. The average of these two values is the fitness function for this calibration. The population in each generation was kept as 50, which was the default if the number of variables is less than five in Optimtool. The error for calibration was found to be 11% whereas the validation error was found to be 14.34%. The τ_m and τ_f values are as shown in Table 5.

Table 5. Parameter Values Calibrated Using a Genetic Algorithm for the Modified SFM

Serial No.	Parameter name	Values	Gender factor $\alpha_{m/f}$
1	τ_m	0.83	0.83
2	τ_f	1.34	1.34

The gender factors were calculated using Equation 14. $\tau_{m/f}$ and $\alpha_{m/f}$ were the same as the value of τ_i is 1. The gender found from this calibration explains the effect of gender on the average pedestrian speed.

Conclusion

Sidewalks are an important pedestrian facility to cater for pedestrian flow. From the literature we have found that the speed of pedestrians varies based on their characteristics, such as age, gender, and so forth. Our analysis also shows variation in speed between male and female pedestrians in both the directions of flow. An ANOVA test was performed to show that gender affects pedestrian speed significantly. Previous studies have modified the social force equations to model particular pedestrian behavior in their study area but have not modeled pedestrian characteristics such as gender. Therefore, in this study we have introduced a factor $\alpha_{m/f}$ to quantify and model the effect of pedestrian gender. We have chosen a base SFM including the driving force, pedestrian–pedestrian interaction force, and pedestrian–boundary interaction force. This model has six parameters to be calibrated, which are the relaxation time τ_i, the pedestrian–pedestrian strength A_i, the pedestrian–pedestrian interaction range B_i, the boundary interaction strength A_{iW}, the boundary interaction range B_{iW}, and λ. A genetic algorithm was used to calibrate the base SFM. Absolute percentage error between the field and simulated speed is the fitness function. Values of the six parameters were found. In the next step two relaxation time parameters were used, one for male pedestrians τ_m and the other for female pedestrians τ_f. Keeping all other parameters as constant based on previous calibration, only the reaction time for male and female pedestrians was calibrated and found. From this gender factor α_m and α_f were calculated. These factors can be used to model the effect of each gender on the overall pedestrian speed.

In this study we have calibrated the base SFM using a genetic algorithm, which is an effective way to calibrate SFMs, especially in the absence of trajectory data. We have also introduced a factor to model the gender effects of pedestrians. The gender factors should be later found for different flows of male and female pedestrians. Similar factors can be found in future studies for pedestrian characteristics such as age and group behavior, so as to help in the design of pedestrian sidewalks for different compositions of gender, age, and groups. The proposed model in this study is applicable for all sidewalks of similar conditions.

References

1. LEE International. *Comprehensive Transportation Study for MMRDA*, Mumbai Metropolitan Region Development Authority, 2008.
2. IRC. *IRC: 103-2012, Guidelines For Pedestrian Facilities*. Indian Road congress, 2012.
3. Laxman, K. K., R. Rastogi, and S. Chandra. Pedestrian Flow Characteristics in Mixed Traffic Conditions. *Journal of Urban Planning and Development*, Vol. 136, No. 1, 2010, pp. 23–33. https://doi.org/10.1061/(ASCE)0733-9488(2010)136:1(23).
4. Morrall, J. F., L. L. Ratnayake, and P. N. Seneviratne. Comparison of Central Business District Pedestrian Characteristics in Canada and Sri Lanka. *Transportation Research Record: Journal of the Transportation Research Board,* 1991. 1294: 57–61.
5. Tarawneh, M. S. Evaluation of Pedestrian Speed in Jordan with Investigation of Some Contributing Factors. *Journal of Safety Research*, Vol. 32, No. 2, 2001, pp. 229–236. https://doi.org/10.1016/S0022-4375(01)00046-9.
6. Finnis, K. K., and D. Walton. Field Observations to Determine the Influence of Population Size, Location and Individual Factors on Pedestrian Walking Speeds. *Ergonomics*, Vol. 51, No. 6, 2008, pp. 827–842. https://doi.org/10.1080/00140130701812147.
7. Tanaboriboon, Y., S. S. Hwa, and C. H. Chor. Pedestrian Characteristics Study in Singapore. *Journal of Transportation Engineering*, Vol. 112, No. 3, 1986, pp. 229–235.
8. Polus, A., J. L. Schofer, and A. Ushpiz. Pedestrian Flow and Level of Service. *Journal of Transportation Engineering*, Vol. 109, No. 1, 1983, pp. 46–56. https://doi.org/10.1061/(ASCE)0733-947X(1983)109:1(46).
9. Koushki, P. A. Walking Characteristics in Central Riyadh, Saudi Arabia. *Journal of Transportation Engineering*, Vol. 114, No. 6, 1989, pp. 735–744.
10. Lewin, K., D. Cartwright, and D. Price. Field Theory in Social Science: Selected Theoretical Papers. *American Sociological Review*, Vol. 16, No. 3, 1951, p. 404. https://doi.org/10.1037/0021-9010.69.1.85.
11. Helbing, D., and P. Molnar. Social Force Model for Pedestrian Dynamics. *Physical Review E*, Vol. 51, No. 5, 1998, pp. 4282–4286.
12. Helbing, D., and I. Farkas. Simulation of Pedestrian Crowds in Normal and Evacuation Situations. *Pedestrian and Evacuation Dynamics*, Vol. 21, No. 2, 2002, pp. 21–58.
13. Helbing, D., and A. Johansson. Pedestrian, Crowd, and Evacuation Dynamics. *Encyclopedia of Complexity and Systems Science*, Vol. 16, No. 4, 2013, pp. 6476–6495. https://doi.org/10.1007/978-3-642-04504-2.
14. Lakoba, T. I. Modifications of the Helbing-Molnar-Farkas-Vicsek Social Force Model for Pedestrian Evolution. *Simulation*, Vol. 81, No. 5, 2005, pp. 339–352. https://doi.org/10.1177/0037549705052772.
15. Shuaib, M. M., O. Moh'd Alia, and Z. Zainuddin. Incorporating Prediction Factor into the Investigation Capability in the Social Force Model: Application on Avoiding Grouped Pedestrians. *Applied Mathematics and Information Sciences*, Vol. 7, No. 1, 2013, pp. 323–331. https://doi.org/10.12785/amis/070141.
16. Zainuddin, Z., and M. M. A. Shuaib. Modelling the Independence Factor and Its Effect on the Preferred Force of the Social Force Model in Emergency and Non-Emergency Situations The Social Force Model. *Applied Mathematics & Information Sciences*, Vol. 5, No. 1, 2011, pp. 53–64.
17. Höcker, M., P. Milbradt, and A. Seyfried. Simulation of Pedestrian Dynamics and Model Adjustments: A Reality-Based Approach. *Proceedings of the Conference on Traffic and Granular Flow*, 2009, pp. 1–6.
18. Hou, L., J. G. Liu, X. Pan, and B. H. Wang. A Social Force Evacuation Model with the Leadership Effect. *Physica A: Statistical Mechanics and its Applications*, Vol. 400, 2014, pp. 93–99. https://doi.org/10.1016/j.physa.2013.12.049.
19. Parisi, D. R., M. Gilman, and H. Moldovan. A Modification of the Social Force Model Can Reproduce Experimental Data of Pedestrian Flows in Normal Conditions. *Physica A: Statistical Mechanics and its Applications*, Vol. 388, No. 17, 2009, pp. 3600–3608. https://doi.org/10.1016/j.physa.2009.05.027.
20. Zhang, H., M. Jin, and H. Liu. A Modified Social Force Model under Different Psychological Conditions for Room Evacuation. In *ICTIS 2013: Improving Multimodal Transportation Systems-Information, Safety, and Integration*, 2013, pp. 1262–1267. https://doi.org/10.1061/9780784413036.169.
21. Xi, H., Y. J. Son, and S. Lee. An Integrated Pedestrian Behavior Model Based on Extended Decision Field Theory and Social Force Model. *Proc., 2010 Winter Simulation Conference (WSC)*, 2010, pp. 824–836. https://doi.org/10.1109/WSC.2010.5679108.
22. Seyfried, A., B. Steffen, and T. Lippert. Basics of Modelling the Pedestrian Flow. *Physica A: Statistical Mechanics and its Applications*, Vol. 368, No. 1, 2006, pp. 232–238. https://doi.org/10.1016/j.physa.2005.11.052.
23. Gao, Y., P. B. Luh, H. Zhang, and T. Chen. *A Modified Social Force Model Considering Relative Velocity of Pedestrians*, 2013 IEEE International Conference on Automation Science and Engineering (CASE), Madison, WI, 2013, pp. 747–751. doi: 10.1109/CoASE.2013.6654008.
24. Kwak, J., H. H. Jo, T. Luttinen, and I. Kosonen. Collective Dynamics of Pedestrians Interacting with Attractions. *Physical*

Review E, Vol. 88, No. 6, 2013, p. 62810. https://doi.org/10.1103/PhysRevE.88.062810.
25. Zainuddin, Z., and M. Shuaib. Modelling the Pedestrian's Perception for the Social Force Model to Reproduce Experimental Data. *Applied Sciences*, Vol. 11, No. 7, 2010, pp. 865–870.
26. Wu, H., J. Huang, Z. L. Guo, and Y. G. Hu. Evacuation Simulation in the Narrow Stair with Revised Social Force Model. *Applied Mechanics and Materials*, Vol. 409–410, 2013, pp. 1577–1582. https://doi.org/10.4028/www.scientific.net/AMM.409-410.1577.
27. Moussaïd, M., N. Perozo, S. Garnier, D. Helbing, and G. Theraulaz. The Walking Behaviour of Pedestrian Social Groups and Its Impact on Crowd Dynamics. *PLoS One*, Vol. 5, No. 4, 2010, p. e10047. https://doi.org/10.1371/journal.pone.0010047.
28. Johansson, A., D. Helbing, and P. K. Shukla. Specification of the Social Force Pedestrian Model by Evolutionary Adjustment to Video Tracking Data. *Advances in Complex Systems*, Vol. 10, No. 2, 2007, pp. 271–288. https://doi.org/10.1142/S0219525907001355.
29. Tang, M., and J. Hongfei. *An Approach for Calibration and Validation of the Social Force Pedestrian Model*, Proceedings 2011 International Conference on Transportation, Mechanical, and Electrical Engineering (TMEE), Changchun, 2011, pp. 2026-2031. doi: 10.1109/TMEE.2011.6199614.
30. Kawamoto, K. A Data Assimilation Method for Estimating the Parameters of a Social Force Model for Pedestrian Motion Analysis. In World Automation Congress, IEEE, 2010, pp. 1–5.
31. Ko, M., T. Kim, and K. Sohn. Calibrating a Social-Force-Based Pedestrian Walking Model Based on Maximum Likelihood Estimation. *Transportation*, Vol. 40, No. 1, 2012, pp. 91–107. https://doi.org/10.1007/s11116-012-9411-z.
32. Zeng, W., P. Chen, H. Nakamura, and M. Iryo-Asano. Application of Social Force Model to Pedestrian Behavior Analysis at Signalized Crosswalk. *Transportation Research Part C: Emerging Technologies*, Vol. 40, 2014, pp. 143–159. https://doi.org/10.1016/j.trc.2014.01.007.
33. EasyFit 5.6. Mathwave Technologies. http://www.mathwave.com/easyfit-distribution-fitting.html, 2017.

The Standing Committee on Transportation in the Developing Countries (ABE90) peer-reviewed this paper (18-05852).

Article

TRR
JOURNAL OF THE TRANSPORTATION RESEARCH BOARD

Transportation Research Record
2018, Vol. 2672(31) 10–20

DOI: 10.1177/0361198118755683
journals.sagepub.com/home/trr
SAGE

Interactions between Driving Skills on Aggressive Driving: Study among Chinese Drivers

Chaozhong Wu[1], Wenhui Chu[1,2], Hui Zhang[1], and Türker Özkan[3]

Aggressive driving has attracted significant attention recently with the increase in related road traffic collisions occurring in China. This study aims to investigate the effect of driving skills on aggressive driving behaviors and traffic accidents to find implications for traffic safety improvement in China. A total of 735 Chinese drivers were recruited to complete a self-reported survey including demographic information, the translated Driver Skill Inventory (DSI), and Driver Aggression Indicator Scale (DAIS). Exploratory factor analysis was first conducted to investigate the factor structures of DSI and DAIS among Chinese drivers. Unlike the two-factor solution (i.e., perceptual-motor and safety skills) found in other studies, the current study result revealed a three-factor solution (i.e., perceptual-motor, safety, and emotional control skills) of DSI. Then, the interaction between DSI factors on DAIS factors, demographic variables, and the number of self-reported traffic accidents and offenses was tested by using moderated regression methods. The results revealed the interaction between perceptual-motor skills and safety skills on aggressive warnings committed by drivers themselves. The interactive effect between safety skills and emotional control skills on perceived aggressive warnings was also found. The results suggested that higher ratings of safety skills are essential for buffering the effect of high-level perceptual-motor skills and emotional control skills on aggressive driving in China. In conclusion, policy makers should be interested in understanding the effect of Chinese drivers' skills on the aggression drivers committed and conceived in traffic. Successful intervention strategies should include all skill factors in the driver training contents.

In recent years, China has witnessed a rapid increase of urbanization and motorization (*1,2*). This has been accompanied by a dramatic change of traffic environment that has led to an increase in driving aggression (*3*) and traffic fatalities (*4,5*). A survey conducted in China indicated that about 60.72% of drivers had experienced aggressive driving (*6*). Considering road traffic deaths per 100,000 population, China has a rate more than double the average of high-income countries globally (18.8 relative to 9.2; see World Health Organization [*4*]). Studies on aggressive driving and traffic accidents could be important in improving traffic safety in China (*7*).

Driving skills were found to be related to aggressive driving and road accident involvement (*8, 9*). Driving skills refer to the ways drivers can drive, in other words, they describe the potential competencies and abilities of drivers (*10*). Prior studies have identified driving skills as a combination of perceptual-motor skills (e.g., fluent lane-changing in heavy traffic) and safety skills (e.g., perceiving hazards in traffic) (*11*). Researchers have indicated that overconfidence in driving skills might make a driver be impatient with other drivers on the road and easily become irritated and aggressive (*12*). Martinussen et al. found that drivers who reported low levels of safety skills reported a high frequency of aggressive driving behaviors (*9*). It has also been proposed that low ratings of safety skills and high ratings of perceptual-motor skills are related to dangerous driving and road accident involvement (*13*).

Moreover, previous studies found an asymmetric interaction between perceptual-motor skills and safety skills when predicting accidents and aggressive driving behaviors (*13, 14*). Sümer et al. tested the interactions between perceptual-motor and safety skills on a number of outcome variables such as traffic accidents, penalties, and aggressive behaviors in Turkey (*14*). They found that perceptual-motor skills

[1]Intelligent Transportation Systems Research Center, Wuhan University of Technology Wuchang District, Wuhan, Hubei, China
[2]School of Automobile and Traffic Engineering, Wuhan University of Science and Technology, Wuchang District, Wuhan, Hubei, China
[3]Safety Research Unit, Department of Psychology, Middle East Technical University, Üniversiteler Mahallesi, Ankara, Turkey

Corresponding Author:
Address correspondence to Hui Zhang: zhanghuiits@whut.edu.cn

moderated the effects of safety skills on six out of eight dependent variables. Based on their findings, it was proposed that overestimating one's perceptual-motor skills should be buffered by a high emphasis on safety skills to avoid risky driving. Özkan et al. investigated the interaction between driving skills on offenses and accidents in six countries (*13*). The asymmetric interaction between perceptual-motor and safety skills on penalties was found in Finland and Turkey. Thus, a better understanding of driving skills could be an effective measure to help reduce aggressive driving behaviors and improve traffic safety.

Few researchers have studied the driving skills of Chinese drivers. Zhang et al. conducted a focus group discussion in China (*15*). Their findings indicated that compared with US drivers, Chinese drivers concentrate more on driving skills and capabilities rather than practical safe driving guidelines. Shi et al. applied a 10-item Driver Skill Inventory (DSI) in China (*16*). Their findings showed that safety skills had a strong impact on reducing aberrant behaviors whereas self-assessed perceptual-motor skills had no impact on aberrant behaviors. Chu et al. compared the effects of driving skills and driving experience on aggressive driving (*17*). The results indicated that it is not Chinese drivers with the highest level of driving skills that commit the highest frequency of aggressive driving behaviors, but those with the highest level of driving experience. It is important to note that the interactions between perceptual-motor and safety skills when predicting aggressive driving behaviors and traffic accidents involvement have not been tested in China. Furthermore, to the best of the authors' knowledge, the interaction between perceptual-motor and safety skills on the perception of other drivers' aggressive behaviors has not been tested.

DSI has been confirmed to have good performance in understanding the driver's perceptual-motor skills and safety skills and reflecting the driver's attitude to safety (*10*, *13*). DSI was developed by Lajunen and Summala (*10*). It is composed of 20 items using five-point Likert scales (1 = definitely weak to 5 = definitely strong), asking drivers to indicate their strong and weak components on driving skills in relation to an internal reference. Several studies have been performed concerning the factorial structure of DSI. The two-factor solution (i.e., perceptual-motor skills and safety skills) has been confirmed to have good reliability in Australia (*11*), the UK (*18*), the Netherlands, Greece, Iran, Turkey, Finland, Sweden, and Denmark (*13*, *14*, *19*, *20*). Shi et al. applied a 10-item version of the DSI in China (*16*). However, the full version of DSI (20 items) has not been tested in China.

As there is evidence that social behavior, cognitive processes, and attitudes are influenced by cultural background (*21*, *22*), the view of Chinese drivers on driving skills and aggressive driving behaviors might be different from other countries. Thus, this study aims to test the factorial structures of the 20-item DSI and Driver Aggression Indicator Scale (DAIS) to develop a deep understanding of Chinese drivers' driving skills and to investigate the relationship between driving skills, aggressive driving behaviors, accidents, and offenses. The study also aims to test the interactive effects between DSI factors on accidents, offenses, and aggressive driving behaviors drivers commit and perceive in China.

The paper is organized as follows. In the next section we describe the measures, participants, and statistical methods. Next, we present the results of this study, including the factor structures of the DSI and DAIS, and the relationship between DSI, DAIS, and involvement in accidents and offenses. This is followed by a discussion.

Methods

Measures

DSI was used to measure driving skills in this study. Aggressive driving behaviors was measured by DAIS developed by Özkan et al. and the reliability and validity of DAIS had been validated in several studies (*23*, *24*). The DAIS contains a "self" part and an "other" part, using a five-point scale from "0 = never" to "4 = nearly all the time." In the "self" part of the scale, drivers were asked to indicate how often they have committed each of the given behaviors on the items listed. In the "other" part of the scale, drivers were asked to indicate how often they perceived those behaviors from other drivers.

The number of traffic accidents and offenses drivers were involved in along with other demographic information such as age, gender, and annual kilometers driven were gathered. Traffic accidents were defined as all types of crashes drivers experienced over the last 3 years. Traffic offenses were defined as penalties paid for illegal parking, excessive speed, running red lights, and other offenses over the past 3 years.

Therefore, a questionnaire including DSI, DAIS, and questions related to background information was used in this study. The measurements were translated into Chinese by two psychologists specializing in driving behavior and back translation was used to ensure the correctness.

Participants and Procedures

A web-based data collection method was used through cooperation with one professional research company in China. The participation was voluntary, but having a valid driving license C1 (in China, people holding a driving license C1 can drive a car or small van) was obligatory. The participants were assured of confidentiality and anonymity of their information. A total of 908 samples were obtained from 27 provinces in China.

Prior the main analyses, a series of preprocessing procedures was carried out on the original data. Considering the recommendations of Elander et al., questions on accident involvement were limited to the last 3 years (*8*) and, thus, 83 drivers with less than 3 years of driving experience were excluded from the sample. In addition, 24 participants with extensive missing data and 44 participants with annual mileages less than 100 km were excluded. Outlier analysis through z-transformation was run on mileage, accidents, and offenses. Cases with z-values less than –3.29 or more than +3.29 were excluded from the analysis. After the data cleaning process, 735 samples representing 81% were found to be appropriate for the study. However, out of the total valid sample, only 479 participants indicated their traffic offenses.

Statistical Analyses

The data were analyzed by using descriptive statistics, factor analyses, Pearson's correlations, and moderated hierarchical regression analysis.

In order to see the driving skill characteristics of Chinese drivers, principal component analysis (PCA) with Promax rotation and Kaiser normalization was carried out on all the items of the DSI scales. The factor structures of DAIS were also tested.

Pearson's correlation was calculated to describe the strength and direction of the linear relationship between DSI factors, DAIS factors of "self" and "other," the number of accidents and offenses as well as background variables.

To investigate the interaction between DSI factors on accidents, offenses, and aggressive driving behaviors, a series of moderated hierarchical regression analyses were run. For the regression, age, gender, and annual mileage were entered in the first step as control variables; DSI scales were entered in the second step as the main effects and their interactions were entered in the third step. Following the procedure outlined by Aiken and West (*25*), moderator and independent variables were first centered (i.e., converted into deviation score form so that their means are zero) and then interaction terms (multiplying together each two centered variables) were created before the analyses. To structure and interpret the complex multiple regression models, significant interactions were plotted by generating simple regression equations of a given dependent (outcome) variable (e.g. aggressive warnings [self]) at low (i.e., one standard deviation below the mean), moderate (mean), and high (i.e., one standard deviation above the mean) levels of DSI factor scores.

Results

Characteristics of the Sample

As presented in Table 1, the characteristics of the study sample included 297 women (40.4%) and 438 men (59.6%), with the age ranging from 22 to 64 years. The majority were between 31 and 40 years representing 51.2% of the participants. The mean of the annual mileage was 12,290 km (standard deviation [SD] = 9.08) with 61.4% of the participants having between one and four traffic accidents over the last three years. A total of 15.2% of the participants reported no offenses and 49.7% had between one and three offenses over the last three years. More details about the characteristics of the participants can be found in Table 1.

Factorial Structure of DSI

The data were subject to PCA to examine the factor loadings and structure. The Bartlett's test of sphericity was significant ($p < .001$) and the Kaiser–Mayer–Olkin (KMO) index was an acceptable value (KMO = 0.932), indicating that the data was appropriate for factor analysis. Different from the previous studies with two-factor structure dominating (*9*, *11*, *13*, *20*), in this current study, a three-factor structure was found to be appropriate for the Chinese driving population. As presented in Table 2, after the Promax rotation, the first factor explained 36.47% of the variance and loaded as perceptual-motor skills with 10 items, which is similar to previous studies. The second factor loaded as safety skill with six items and 11.98% of the variance was explained. The third factor, which is the newly found structure among Chinese drivers, consisted of three items reflecting emotional control ability toward safety. Thus, the third factor was named as emotional control skills. The three factors together accounted for 54.71% of the variance. However, one item, "Controlling the vehicle," was excluded from the subsequent analysis as it cross-loaded on two factors. The reliability coefficients were acceptable. The means and standard deviations for DSI items and the factor loadings are presented in Table 2.

Factorial Structure of DAIS

Similarly, PCAs were carried out on the 13 items of DAIS for both "self" and "other" parts. Both Cattell's scree plot and eigenvalues indicated that a two-factor solution fits better and the two parts of DAIS shared the same structure. The items in the first factor reflected aggressive warnings on the road (e.g., sounded horn). It consisted of seven items and was labeled as "aggressive warnings." The second factor consisted of six items (e.g., physically attacked) and was labeled as "hostile aggression and revenge." The reliability coefficients were considered acceptable. The means and standard deviations for DAIS items and the factor loadings are shown in Table 3.

Correlations between DSI, DAIS Factors, and Background Variables

Correlations among background variables (age, gender, annual mileage, the number of accidents, and offenses), DSI factors, and DAIS factors of "self" and "other" parts in the total sample are presented in Table 4.

In Chinese samples, age was negatively related to emotional control skills. Gender correlated positively with

Table 1. Sample Characteristics

Variables	N	Mean/SD	Percentage (%)
Age			
Total	735	34.97/7.22	
Male	438	36.15/7.88	59.6
Female	297	33.23/5.70	40.4
22–30 years old	212	-	28.8
31–40 years old	376	-	51.2
41–64 years old	147	-	20.0
Annual mileage (×1,000 km)			
Total	-	12.29/9.08	-
Male	-	13.12/9.02	-
Female	-	11.06/9.06	-
≤5	165	-	22.5
5.001–10	248	-	33.7
10.001–20	250	-	34.0
≥20.001	72	-	9.8
Accidents involvement			
Total	na	2.36/2.18	na
Male	na	2.51/2.26	na
Female	na	2.14/2.05	na
None	165	na	22.4
1–4	451	na	61.4
5–6	89	na	12.1
≥7	30	na	4.1
Number of traffic offenses			
Total	479	3.16/2.75	
Male	289	3.24/2.84	60.3
Female	190	3.04/2.61	39.7
None	73	na	15.2
1-3	238	na	49.7
4-6	110	na	15.0
7-9	44	na	9.2
≥10	14	na	1.9

Note: na = not applicable.

perceptual-motor skills and number of accidents and negatively correlated with emotional control skills. Annual mileage correlated positively with perceptual-motor skills.

Perceptual-motor skills were negatively related to aggressive warnings on the "self" and "other" part. Both safety skills and emotional control skills were negatively associated with aggressive warnings and hostile aggression and revenge behaviors on both "self" and "other" parts of the scale. As expected, both DSI and DAIS factors were significantly and positively related to each other.

Perceptual-motor skills were not significantly associated with the number of traffic accidents and offenses. On the other hand, safety skills were negatively related to the number of traffic offenses, but had no significant correlation with the number of traffic accidents. Emotional control skills were negatively associated with the number of traffic accidents.

Moderated Hierarchical Regression Analyses

As presented in Table 5, control variables in the first step were significantly related to the outcome variables except for the traffic offenses. Gender was significantly associated with the number of accidents and aggressive driving behaviors perceived and committed. Annual mileage was significantly associated with aggressive warnings of both "self" and "other" scale.

After controlling the effects of the background variables, perceptual-motor skills, safety skills, and emotional control skills were consistently and significantly related to all the outcome variables. As indicated in Table 5, perceptual-motor skills were significantly associated with the number of accidents ($\beta = -0.09$, $p < .05$), hostile aggression and revenge (self) ($\beta = 0.16$, $p < .001$) and aggressive warnings (other) ($\beta = -0.27$, $p < .001$). The safety skills were significantly related to the number of traffic offenses ($\beta = -0.16$, p

Table 2. Means and Standard Deviations for DSI Items and the Three-Factor Solution

Items (code)	Mean/SD	Perceptual-motor skills	Safety skills	Emotional control skills
Overtaking (14)	3.46/.85	.84		
Fluent lane-changing in heavy traffic (7)	3.62/.94	.84		
Knowing how to act in particular traffic situations (6)	3.74/.79	.76		
Managing the car through a skid (4)	3.59/.82	.73		
Making firm decisions (8)	3.90/78	.71		
Make a hill start on a steep incline (13)	3.87/.81	.69		
Reverse parking into a narrow gap (20)	3.83/.84	.66		
Predicting traffic situations ahead (5)	3.92/.73	.61		
Perceiving hazards in traffic (2)	3.93/.73	.56		
Fluent driving (1)	4.10/.67	.55	.35	
Conforming to the speed limits (16)	4.17/.72		.81	
Obeying the traffic lights carefully (19)	4.32/.68		.80	
Keeping a sufficient following distance (11)	4.11/.72		.70	
Avoiding unnecessary risks (17)	4.13/.70		.64	
"Relinquishing" legitimate rights when necessary (15)	4.08/.72		.52	
Adjusting your speed to the conditions (12)	4.09/.70	.36	.50	
Controlling the vehicle (10)*	4.09/.66	.44	.45	
Driving behind a slow car without getting impatient (3)	3.72/.97			.83
Tolerating other drivers' errors calmly (18)	3.77/.84			.77
Staying calm in irritating situations (9)	3.85/.83			.76
Mean/SD		3.79/.56	4.15/.50	3.78/.73
Initial Eigenvalues		7.30	2.40	1.25
Variance Explained (%)		36.47	11.98	6.25
Cronbach's Alpha		.89	.80	.78

Note: Loadings less than 0.3 were suppressed and 0.36 was used as a cutoff point for a better interpreting of factors. Items with an asterisk were subsequently removed from the scale to optimize the psychometric properties of the DSI. KMO was 0.932; Bartlett sigma <0.001.

Table 3. Means and Standard Deviations for DAIS Items and the Two-Factor Solution

	Aggressive Behaviors (self)			Aggressive Behaviors (other)		
Items(Code)	Mean/SD	Aggressive warnings (AW)	Hostile aggression and revenge (HAR)	Mean/SD	Aggressive warnings (AW)	Hostile aggression and revenge (HAR)
Rammed a vehicle (5)	.68/.74	.75		1.71/1.07	.81	
Cut up (1)	.67/.68	.77		1.86/1.00	.78	
Prevented or obstructed from maneuvering the vehicle (11)	.44/.62	.69		1.08/.93	.78	
Drove slowly in order to annoy the driver behind (12)	.65/.73	.57		1.23/.91	.76	
Hugged the rear bumper (10)	.60/.70	.71		1.25/.91	.71	
Sounded horn. (13)	.50/.64	.77		1.37/.99	.71	
Flashed lights (8)	.51/.70	.67		1.20/.91	.71	
Physically attacked (6)	.12/.44		.91	.27/.56		.89
Threatened physically (9)	.11/.36		.86	.30/.61		.86
Made a hand gesture (3)	.23/.49		.67	.60/.76		.79
Threatened verbally (4)	.18/.45		.73	.59/.77		.77
Swore/verbally abused (2)	.35/.59		.45	.82/.82		.69
Chased (7)	.48/.68		.38	.70/.86	.36	.46
Initial Eigenvalues		5.39	1.40		6.02	1.68
Cronbach's α values		.84	.79		.87	.86
Variance (%)		41.46	10.77		46.31	12.91

Note: Factor loadings below 0.30 were omitted for clarity and 0.36 was used as a cutoff point for a better interpreting of factors. For the "other" part, KMO was 0.918, Bartlett sigma <0.001. For the "self" part, KMO was 0.906, Bartlett sigma <0.001.

Table 4. Correlations among Background Variables, DSI Factors, and DAIS Factors

	1	2	3	4	5	6	7	8	9	10	11
1. Age	-										
2. Gender[a]	.20***	-									
3. Annual mileage	−.02	.11**	-								
4. Accidents involvement	−.03	.08*	.06	-							
5. Traffic Offenses[b]	−.05	.04	.05	.40***	-						
6. P-M skills	−.03	.25***	.19***	−.07	−.07	-					
7. SS	−.06	−.04	.03	−.07	−.13**	.51***	-				
8. ECS	−.08*	−.15***	.04	−.11***	−.08	.31***	.52***	-			
9. AW (self)	.01	.16***	.11***	.27***	.34***	−.11***	−.25***	−.31***	-		
10. HAR (self)	−.01	.14***	.09*	.12***	.20***	−.01	−.31***	−.20***	.61***	-	
11. AW (other)	.02	.07	.03	.24***	.21***	−.21***	−.10***	−.22***	.42***	.11***	-
12. HAR (other)	.02	.10***	.06	.23***	.26***	−.05	−.17***	−.10**	.42***	.47***	.59***

Note: Missing values were excluded pairwise for the analysis. P-M skills = perceptual-motor skills; SS = safety skills; ECS = emotional control skills; AW (self) = aggressive warnings (self); HAR (self) = hostile aggression and revenge (self); AW (other) = aggressive warnings (other); HAR (other) = hostile aggression and revenge (other).
[a]Gender: 1, female; 2, male.
[b]Correlation between offenses and other variables was calculated based on 479 samples.
*$p < .05$, **$p < .01$, ***$p < .001$.

< .05) and aggressive warnings and hostile aggression and revenge of both "self" and "other" scale. Furthermore, the emotional control skills were negatively related to both "self" ($\beta = -0.22, p < .001$) and "other" ($\beta = -0.20, p < .001$) aggressive warning. In the third step, different from the previous study (*14*), the interactions between perceptual-motor skills, emotional control skills, and safety skills predicted two of the six outcome variables (aggressive warnings of "self" and "other").

The significant predictors together accounted for 4% of the variance in both traffic accident and offenses. In addition, the total variance explained in aggressive warning "self" and "others" as well as hostile aggressive revenge "self" and "others" were 14%, 11%, 15%, and 5%, respectively.

For a better understanding of the interactive effects of DSI factors on aggressive driving behaviors, significant interactions between perceptual-motor skills, emotional control skills, and safety skills were plotted.

As shown in Figure 1*a*, among Chinese drivers with a low ($t(731) = 3.69, p < .01$), moderate ($t(731) = 6.15, p < .01$), and high ($t(731) = 6.5, p < .01$) level of perceptual-motor skills, the ratings of aggressive warnings committed by drivers themselves decreased as their safety skills increased. As demonstrated, drivers with high ratings of perceptual-motor skills and low ratings of safety skills reported the highest level of aggressive warning behaviors, and those with high ratings of perceptual-motor skills and high ratings of safety skills reported the lowest level of aggressive warning behaviors. High ratings in safety skills buffered the perceptual-motor skills' effect on aggressive warnings (self).

Figure 1*b* clearly shows that among Chinese drivers with a low ($t(731) = 2.41, p < .05$), moderate ($t(731) = 2.55, p < .05$), and high ($t(731) = 2.13, p < .05$) level of emotional skills, the ratings of aggressive warnings committed by drivers themselves decreased as their safety skills increased. In other words, drivers with low ratings of emotional control skills and low ratings of safety skills reported the highest level of aggressive warning behaviors, and those with high ratings of emotional control skills and high ratings of safety skills reported the lowest level of aggressive warning behaviors.

From Figure 1*c*, among drivers with low ($t(731) = 5.48, p < .01$), moderate ($t(731) = 5.03, p < .01$), and high ($t(731) = 3.10, p < .01$) ratings of safety skills, their perceived aggressive warnings decreased as the emotional control skills increased. As can be seen, drivers with low ratings of emotional control skills and low ratings of safety skills reported to perceive the highest level of aggressive warnings from other drivers, and those with high ratings of emotional control skills and low ratings of safety skills reported to perceive the lowest level of aggressive warnings from other drivers. High ratings of safety skills buffered the effect of emotional control skills' on aggressive warnings (other).

Discussion

With the aim of investigating the driving skill characteristics of Chinese drivers and to test the interaction between driving skills on aggressive driving and accidents involvement, this study applied the 20-item DSI and DAIS among Chinese drivers. The factorial structure of DSI and DAIS was assessed and the relationship between driving skills, aggressive driving behaviors, accidents, and offenses was investigated. Finally, interaction effects between DSI factors on the number of accidents, traffic offenses, and aggressive driving behaviors were tested and plotted.

Table 5. Moderated Hierarchical Regression Analyses on Outcome Variables

	DV: Accidents involvement			DV: Traffic Offenses			DV: Aggressive warnings (self)			DV: Hostile aggression and revenge(self)			DV: Aggressive warnings (other)			DV: Hostile aggression and revenge(other)		
	R^2 Change	F	Beta	R^2 Change	F	Beta	R^2 Change	F	Beta	R^2 Change	F	Beta	R^2 Change	F	Beta	R^2 Change	F	Beta
1. Step: background variables	.01	2.90*	na	.01	1.05	na	.03	8.39***	na	.03	6.20***	na	.01	1.24	na	.01	3.07*	na
Age	na	na	−.07	na	na	−.07	na	na	−.05	na	na	−.06	na	na	−.01	na	na	−.01
Gender[a]	na	na	.10**	na	na	.05	na	na	.13**	na	na	.08*	na	na	.10*	na	na	.09*
Annual mileage	na	na	.06	na	na	.05	na	na	.11**	na	na	.06	na	na	.08*	na	na	.05
2. Step: main effects	.02	3.95**	na	.02	2.76*	na	.10	27.64***	na	.11	31.60***	na	.09	23.89***	na	.03	6.86***	-
PMS	na	na	−.09*	na	na	−.03	na	na	−.05	na	na	.16***	na	na	−.27***	na	na	.00
SS	na	na	−.01	na	na	−.16*	na	na	−.11*	na	na	−.39***	na	na	.17***	na	na	−.14**
ECS	na	na	−.08	na	na	−.01	na	na	−.22***	na	na	−.06	na	na	−.20***	na	na	.00
3. Step: interactions	.01	2.59	na	.01	2.31		.01	4.07**	na	.01	2.55	na	.01	3.00*	na	.01	1.21	-
PMS×SS	na	na	−.09	na	na	−.04	na	na	−.12*	na	na	−.05	na	na	.05	na	na	.05
PMS×ECS	na	na	−.00	na	na	−.03	na	na	−.04	na	na	−.03	na	na	−.04	na	na	−.03
SS×ECS	na	na	−.03	na	na	−.08	na	na	.09*	na	na	−.03	na	na	.10*	na	na	.06

Note: PMS = perceptual-motor skills; SS = safety skills; ECS = emotional control skills; na = not applicable.
[a]Gender: 1, female; 2, male.
$^{*}p < .05$, $^{**}p < .01$, $^{***}p < .001$.

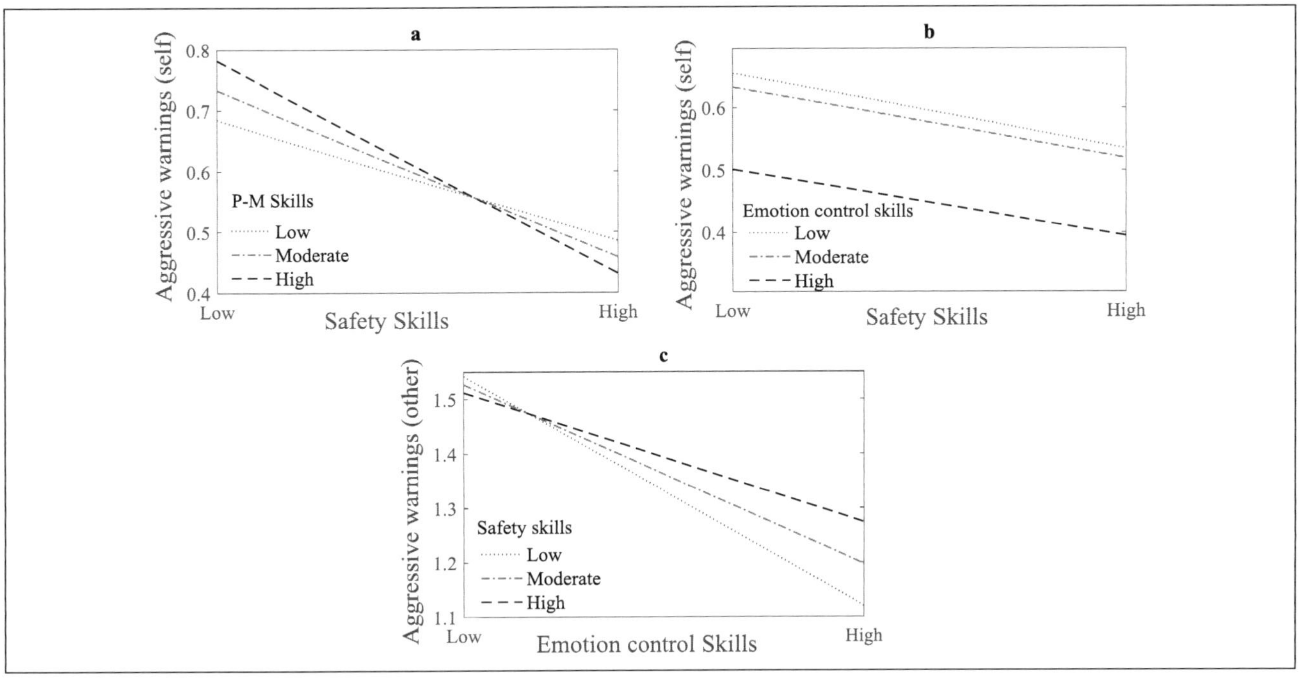

Figure 1. The Interactions between DSI Factors on DAIS Factors.
(*a*) The interaction between safety skills and perceptual-motor skills on aggressive warnings (self) among Chinese samples (P-M skills refers to perceptual-motor skills). (*b*) The interaction between safety skills and emotional control skills on aggressive warnings (self) among Chinese samples. (*c*) The interaction between safety skills and emotional control skills on aggressive warnings (other) among Chinese samples.

Factorial Structures of DSI and DAIS

The result of the exploratory factor analysis revealed that among Chinese drivers, the DSI has a three-factor structure that is different from that found in other countries. This finding confirms the earlier submission that, due to the differences in traffic safety culture among countries, results might vary. The first factor, perceptual-motor skills shared the same contents with previous studies, suggesting that drivers in China have the same requirements of vehicle handling as other parts of the world. This result confirmed the findings of Özkan et al. in their cross-culture study that perceptual-motor skills have a "pan-cultural" characteristic (*13*). The remaining two factors were suggested to be a split of the safety skills factor of earlier studies (e.g., [*13*, *20*]). The safety skills factor included aspects of social tolerance and traffic rule compliance that conformed to the findings of Özkan et al. (*13*). The new third factor (emotional control skills) discovered among Chinese drivers included three items, i.e., "Driving behind a slow car without getting impatient", "Tolerating other drivers' errors calmly," and "Staying calm in irritating situations," all reflecting social tolerance and the control of negative cognitions and emotions while driving. As a dimension of the Dula Dangerous Driving Index, negative cognitions and emotions refer to driving with feelings of anger, frustration, provocation, and irritation (*26*–*28*), and might lead to risky decisions and accidents (*29*). In this study, the negative correlation between emotional control skills, aggressive driving, and accidents might indicate that a high level of emotional control skills can help reducing aggressive driving behaviors and number of accidents.

The finding of a new factor and its negative association with accident and traffic offenses may indicate that Chinese drivers have social tolerance and emotional control necessary for safe driving. It is suggested by Blockey and Hartley that one possible reason for the deviant factorial structure would be differences in the characteristics of the populations studied (*29*). Researchers indicated that China has a short history of automobile usage and an immature traffic management strategy between 2010 and 2014, and presents a unique social and traffic environment (*30*). Again, in China Road Traffic Accidents Statistics (CRTAS), it is reported that risky and illegal driving behaviors accounted for 95.49% of accidents in China (*31*). The traffic environment and culture make it seems reasonable for Chinese drivers to need more emotional control skills while driving.

Consistent with previous studies in different countries, two-factor solution (i.e., aggressive warnings and hostile aggression and revenge) was found for the "self" and "other" part of DAIS (*24*). It should be noted, however, that the content of the two factors of "self" and "other" parts vary from that of the other countries (*24*). For example, "rammed a vehicle" and "cut up" items of "self" and "other" parts of DAIS were perceived as aggressive warnings by Chinese drivers, whereas they perceived "make a hand gesture" item of "self" and "other" parts of DAIS as hostile aggression and revenge. In contrast to previous studies, it seems that Chinese drivers are consistent with the interpretation of aggressive driving behaviors for themselves and others. However, the content of aggressive warnings and hostile aggression and revenge factors for the "self" and "other" parts are the same. This can be used to reduce the possible interpersonal conflicts in traffic.

Relationship between Background Variables, DSI, and DAIS Factors

In Chinese samples, age was negatively related to emotional control skills, indicating that young drivers are facing more social tolerance and negative emotion control problems. Annual mileage correlated positively with perceptual-motor skills and aggressive driving behaviors committed by drivers, indicating that drivers with higher exposure tend to drive aggressively in China. As indicated by Näätänen and Summala, increasing driving experience and exposure to traffic increases the level of driving skills, which, in turn, decreases the sense of subjective risk and concern for safety aspects while driving (*32*).

This result of the relationship between driving skills, aggressive driving behaviors, accidents, and offenses is in line with the studies of Lajunen and Summala (*10*) and Sümer et al. (*14*), demonstrating that perceptual-motor skills have less contribution to accidents and safety skills and emotional control skills may reduce risky driving behaviors among Chinese drivers. In addition, perceptual-motor skills correlated negatively with aggressive warnings on the "self" and "other" part. The moderated regression analyses also revealed that perceptual-motor skills were negatively related to the number of accidents. In general, the findings imply that some level of perceptual-motor skills can have a significant effect on the avoidance of accidents and aggressive driving behaviors in China. It can, therefore, be suggested that, in the Chinese driving setting, perceptual-motor skills can reduce traffic accidents and risky driving behaviors. However, the perceptual-motor skills can be enhanced by supporting it with appropriate driver training, paying more attention to strengthening safety attitude and social tolerance.

The results further revealed that perceptual-motor skills, safety skills and emotional control skills have a positive and significant relation to each other. It appears that perceptual-motor, safety, and emotional control skills can support each other.

Interactions between DSI Factors on Aggressive Driving Behaviors

Finally, the moderation effects of perceptual-motor, safety, and emotional control skills on accidents, offenses and aggressive driving behaviors were tested. The significant interaction between perceptual-motor and safety skills on aggressive warnings of the "self" part was found. Plotting of the interaction revealed that the relationship between safety skills and aggressive warnings (self) varied for drivers with different levels of perceptual-motor skills. This result supports the findings of Özkan et al. (*13*) and Sümer et al. (*14*) that there is an asymmetric relationship between perceptual-motor skills and safety skills when predicting risky driving behaviors.

The significant interaction between emotional control skills and safety skills on aggressive warnings (other) was found. There is an interaction relationship between emotional control skills and safety skills when predicting perceived aggressive behaviors of drivers. Drivers with high ratings of emotional control skills perceive warnings from other drivers as less aggressive, but high ratings of safety skills will make them perceive warnings as more aggressive. The finding implies that in China, drivers with high scores of safety skills are more vulnerable in traffic.

However, there were no significant interactions effects between DSI factors on traffic accidents and offenses. In support of the previous findings (*8*, *13*), this could be that because traffic accidents and offenses are rare events and statistically complex.

Contributions and Conclusion of the Study

To our knowledge, this is the first study to examine the factor structure and to test the validity of the DSI and DAIS on a large sample of drivers in China. Furthermore, the study is the first to our knowledge to test the interaction effects of DSI factors on driver's aggressive behavior and the number of traffic accident involvement among Chinese drivers. The findings of the study have enhanced the understanding of policy makers on the view of Chinese drivers' driving skills for designing remedial measures for modification of aberrant behaviors and improving road safety. In addition, the study has validated the cross-cultural applicability of the DSI and DAIS in China.

In conclusion, among Chinese drivers, DSI factors and their interactions were validated and significant correlation with aggressive driving behaviors and accidents involvement was found. From the findings, it is commendable for the road traffic safety policy makers to have a keen interest in understanding how Chinese drivers perceive themselves and others' driver's aggression in traffic in order to improve the driving and safety skills (including all DSI components). These intervention strategies can be successful by including such skill factors in driver training content.

Limitations and Future Work

Whereas this study has successfully investigated the Chinese drivers' driving skills and its interactions on aggressive driving and accidents involvement among 735 drivers, the samples recruited might be small considering the population of China. Though, the study findings have revealed the possible aberrant behaviors and factors contributing to road traffic crashes and aggressive driving in China for the development of countermeasures. Secondly, the current study is limited by its reliance on drivers' self-reports measure of driving behaviors and traffic accidents as this approach could be influenced by social desirability. However, the anonymous and confidentiality of the questionnaire hopefully might minimize this potential.

Future research should consider recruiting a larger sample of drivers and taking more demographic characteristics (e.g., profession, socio-economic characteristics, education levels, married or unmarried) into account. It is also recommended that field experiments obtained objective observations can be conducted to investigate the relationship between driving skills and aggressive driving behaviors.

Acknowledgments

This study was supported by National Natural Science Foundation of China (u1624262, 61603282), National Key R&D Program of China (Grant No. 2017YFC0804802), and Foundation of Wuhan University of Science and Technology (2016xz028).

References

1. Atchley, P., J. Shi, and T. Yamamoto. Cultural Foundations of Safety Culture: A Comparison of Traffic Safety Culture in China, Japan and the United States. *Transportation Research Part F: Traffic Psychology and Behaviour*, Vol.26, No. PB, 2014, pp. 317–325. https://doi.org/10.1016/j.trf.2014.01.004.
2. Feng, Z., J. Zhan, J. Liu, W. Zhang, and K. Wang. Development and Validation of the Highway Emergency Lane Illegal Occupancy Scale for Chinese Drivers. *Transportation Research Record: Journal of the Transportation Research Board*, 2017. 2634: 43–49.
3. Ding, J. Y. and Lecturer. Study on the Aggressive Driving Behavior Based on Theory of Planned Behavior. *China Safety Science Journal*, Vol. 16, 2006, pp. 15-18.
4. World Health Organization. *Global Status Report on Road Safety 2015*. Geneva: WHO, 2015,
5. Wei, H. Study on Factors Determining Driving Safety in China and the US. *China Safety Science Journal*, Vol. 14, No. 9, 2004, pp. 24–28.
6. Hu, L.. *Characteristics of Angry Driving Behaviors and Its Effects on Traffic Safety*. Wuhan University of Technology, 2011.
7. Kim, D.-G., C. Lee, and B.-J. Park. Use of Digital Tachograph Data to Provide Traffic Safety Education and Evaluate Effects on Bus Driver Behavior. *Transportation Research Record: Journal of the Transportation Research Board*, 2016. 2585: 77–84.
8. Elander, J., R. West, and D. French. Behavioral Correlates of Individual Differences in Road Traffic Crash Risk: An Examination of Methods and Findings. *Psychological Bulletin*, Vol. 113, No. 2, 1993, pp. 279–294.
9. Martinussen, L. M., M. Møller, and C. G. Prato. Assessing the Relationship between the Driver Behavior Questionnaire and the Driver Skill Inventory: Revealing Sub-Groups of Drivers. *Transportation Research Part F: Traffic Psychology and Behaviour*, Vol. 26, Part A, 2014, pp. 82–91.
10. Lajunen, T., and H. Summala. Driving Experience, Personality, and Skill and Safety-Motive Dimensions in Drivers' Self-Assessments. *Personality and Individual Differences*, Vol. 19, No. 3, 1995, pp. 307–318.
11. Lajunen, T., A. Corry, H. Summala, and L. Hartley. Cross-Cultural Differences in Drivers' Self-Assessments of Their Perceptual-Motor and Safety Skills: Australians and Finns. *Personality and Individual Differences*, Vol. 24, No. 4, 1998, pp. 539–550.
12. Lajunen, T., and H. Summala. Effects of Driving Experience, Personality, Driver's Skill and Safety Orientation on Speed Regulation and Accidents. *Nature*, Vol. 198, No. 4876, 1997, pp. 205–206.
13. Özkan, T., T. Lajunen, J. El Chliaoutakis, D. Parker, and H. Summala. Cross-Cultural Differences in Driving Skills: A Comparison of Six Countries. *Accident Analysis and Prevention*, Vol. 38, No. 5, 2006, pp. 1011–1018. https://doi.org/10.1016/j.aap.2006.04.006.
14. Sümer, N., T. Özkan, and T. Lajunen. Asymmetric Relationship between Driving and Safety Skills. *Accident Analysis and Prevention*, Vol. 38, No. 4, 2006, pp. 703–711. https://doi.org/10.1016/j.aap.2005.12.016.
15. Zhang, W., Y. H. Huang, M. Roetting, Y. Wang, and H. Wei. Driver's Views and Behaviors about Safety in China - What Do They NOT Know about Driving? *Accident Analysis and Prevention*. 1. Volume 38, 22–27.
16. Shi, J., Y. Bai, L. Tao, and P. Atchley. A Model of Beijing Drivers' Scrambling Behaviors. *Accident Analysis and Prevention*, Vol. 43, No. 4, 2011, pp. 1540–1546. https://doi.org/10.1016/j.aap.2011.03.008.
17. Chu, W., C. Wu, H. Zhang, C. Atombo, and L. Zhang. Investigating the Relationship Between Driving Skills, Driving Experience and Aggressive Driving Behaviors in China. Presented at 96th Annual Meeting of the Transportation Research Board, Washington, D.C., 2017.
18. Lajunen, T., D. Parker, and S. G. Stradling. Dimensions of Driver Anger, Aggressive and Highway Code Violations and Their Mediation by Safety Orientation in UK Drivers. *Transportation Research Part F: Traffic Psychology and Behaviour*, Vol. 1, No. 2, 1998, pp. 107–121. https://doi.org/10.1016/S1369-8478(98)00009-6.
19. Sümer, N., and T. Özkan. The Role of Driver Behavior, Skills, and Personality Traits in Traffic Accidents. *TURK PSIKOLOJI DERGISI*, Vol. 17, No. 50, 2002, pp. 1–25.
20. Warner, H. W., T. Özkan, T. Lajunen, and G. S. Tzamaloukas. Cross-Cultural Comparison of Driving Skills among Students in Four Different Countries. *Safety Science*, Vol. 57, 2013, pp. 69–74. https://doi.org/10.1016/j.ssci.2013.01.003.
21. Berry, J. W., Y. H. Poortinga, M. H. Segall, and P. R. Dasen. *Cross-Cultural Psychology: Research and Applications*. 2002.
22. Islam, M. T., L. Thue, and J. Grekul. Understanding Traffic Safety Culture: Implications for Increasing Traffic Safety.

Transportation Research Record: Journal of the Transportation Research Board, 2017. 2635: 79–89.
23. Özkan, T., and T. Lajunen. A New Addition to DBQ: Positive Driver Behaviours Scale. *Transportation Research Part F: Traffic Psychology and Behaviour*, Vol. 8, No. 4–5, 2005, pp. 355–368. https://doi.org/10.1016/j.trf.2005.04.018.
24. Özkan, T., T. Lajunen, D. Parker, N. Sümer, and H. Summala. Symmetric Relationship between Self and Others in Aggressive Driving across Gender and Countries. *Traffic Injury Prevention*, Vol. 11, No. November 2012, 2010, pp. 228–239. https://doi.org/10.1080/15389581003788864.
25. Aiken, L. S., and S. G. West. *Multiple Regression: Testing and Interpreting Interactions*. 1991.
26. Dula, C. S., and M. E. Ballard. Development and Evaluation of a Measure of Dangerous, Aggressive, Negative Emotional, and Risky Driving 1. *Journal of Applied Social Psychology*, Vol. 33, No. 2, 2003, pp. 263–282.
27. Iliescu, D., and P. Sârbescu. The Relationship of Dangerous Driving with Traffic Offenses: A Study on an Adapted Measure of Dangerous Driving. *Accident Analysis and Prevention*, Vol. 51, 2013, pp. 33–41. https://doi.org/10.1016/j.aap.2012.10.014.
28. Qu, W., Y. Ge, C. Jiang, F. Du, and K. Zhang. The Dula Dangerous Driving Index in China: An Investigation of Reliability and Validity. *Accident Analysis and Prevention*, Vol. 64, 2014, pp. 62–68. https://doi.org/10.1016/j.aap.2013.11.004.
29. Blockey, P., and L. Hartley. Aberrant Driving Behaviour: Errors and Violations. *Ergonomics*, Vol. 38, No. 9, 1995, pp. 1759–71. https://doi.org/10.1080/00140139508925225.
30. Zhang, W., O. Tsimhoni, M. Sivak, and M. J. Flannagan. Road Safety in China: Analysis of Current Challenges. *Journal of Safety Research*, Vol. 41, No. 1, 2010, pp. 25–30. https://doi.org/10.1016/j.jsr.2009.12.003.
31. Traffic Administration, the M. of P. S. *Statistics of Road Traffic Accidents in P.R. of China*. 2015.
32. Näätänen, R., and H. Summala. *Road-User Behaviour and Traffic Accidents*. 1976.

The Standing Committee on Transportation in the Developing Countries (ABE90) peer-reviewed this paper (18-01174).

Article

TRR
JOURNAL OF THE TRANSPORTATION RESEARCH BOARD

Cyclist Satisfaction Evaluation Model for Free-Floating Bike-Sharing System: A Case Study of Shanghai

Transportation Research Record
2018, Vol. 2672(31) 21–32
© National Academy of Sciences: Transportation Research Board 2018
Article reuse guidelines: sagepub.com/journals-permissions
DOI: 10.1177/0361198118770193
journals.sagepub.com/home/trr
SAGE

Feifei Xin[1], Yifan Chen[1], Xiaobo Wang[2], and Xiaohong Chen[1]

Abstract

The free-floating bike-sharing system (FFBS) is an innovative public bicycle service, which is widely popular throughout China. In comparison to traditional station-based system, FFBS allows cyclists to park the bicycles in any non-motorized parking area so that the convenience can be improved greatly. Moreover, smart devices inside the free-floating bicycle can provide real-time tracking information which offers convenience for intelligent management. This study adopts the structural equation model to measure what factors influence people's willingness to use the system and how the indicators inside the system interact with each other. Unlike the long-standing use of maximum likelihood estimation method, an approach of partial least squares estimation is used in this paper to remove the restriction of multivariate normal distribution. Based on the questionnaire carried out in 32 locations in Shanghai, a conceptual framework cyclist satisfaction index model is established, which includes three levels with 50 indicators. Then, to analyze the different perception among different kinds of cyclists, the study divides them into two groups: commuters and non-commuters. The parameter calibration for commuters and non-commuters are calculated separately. Finally, by virtue of importance-performance analysis (IPA) matrices, the problems in the existing service system are identified. The results demonstrate five aspects including riding environment, riding safety, government intervention, laws and staff service need urgent improvement. According to these findings, innovative countermeasures such as monitoring platform and electronic fences are under construction to prompt the use of FFBS.

The bike-sharing system originated in the Netherlands and was put into practice worldwide for its advantages such as convenience, flexibility and low carbon emission (*1*). However, traditional systems have a major drawback: that is, cyclists must rent and return bicycles in fixed rental stations which reduces the efficiency of the system. As a result, in 2014, the first FFBS company, OFO, was established in China. This innovative way of bicycle use avoids construction of expensive docking stations and kiosk machines. In addition, FFBS can truly realize the seamless connection between public transit and the bike-sharing system because people can park the bicycles in any non-motor vehicle parking area rather than waste time finding docking stations which may be far from rail transit stations or bus stops. As a result, FFBS rapidly gained popularity in China and soon reached a massive scale. Now, nearly forty cities in the country have rolled out FFBS, with twenty million users.

Introduction of the FFBS in the City of Shanghai

There are more than 12 FFBS companies who have launched over 1 million free-floating bicycles in Shanghai city. Two of the biggest enterprises are OFO and Mobike. Smart locks are installed inside the free-floating bicycles and cyclists scan the QR codes on the bicycles with their mobile phones so that the locks will open automatically. The rent fare is charged by the hour but frequent preferential benefits can help to reduce the price. Users pay a deposit before joining the system.

Research Motivation and Objectives

Despite the popularity of FFBS, users' feedback and their degree of satisfaction have not been analyzed in detail. As a result, the main objectives of this study are the following:

- It analyzes the relationship between cyclist satisfaction, perceived quality, cyclist loyalty and complaints, and users' socioeconomic and current usage characteristics by the model's path coefficients.

[1]Key Laboratory of Road and Traffic Engineering of the Ministry of Education, Tongji University, Shanghai, P. R. of China
[2]School of Economics & Management, Tongji University, Shanghai, P. R. of China

Corresponding Author:
Address correspondence to Xiaobo Wang: alex829@tongji.edu.cn

- It discusses the weights of indicators according to the loading coefficients of their corresponding latent variables. Then, the factors which influence the latent variables the most can be identified.
- It allows a comparison to be made between commuters and non-commuters with regard to their different perceptions of the service quality.
- The findings of this study can help to identify FFBS operation and management strategies in order to retain and attract people to use the system.

Literature Review

The degree of customer satisfaction is a key influence on the profits of companies. Many researchers have developed customer satisfaction theories and methodologies. Cronin and Taylor (*3*) analyzed the relationship between customer satisfaction, service quality and purchase intentions in several kinds of service industries. They found that service quality was an antecedent of customer satisfaction and customer satisfaction had a significant effect on purchasing power. Boulding et al. (*4*) showed that there's a strong connection between customer satisfaction and loyalty. Based on prior studies, Eugene et al. (*5*) came up with a customer satisfaction index (CSI) based on their findings from Swedish firms. In 1996, Fornell further improved the CSI model, taking into consideration six variables, customer expectations, perceived quality, perceived value, overall customer satisfaction, customer complaints and customer loyalty, and renamed it the American customer satisfaction index (ACSI) (*6*). In the ACSI model, perceived quality and customer expectations are regarded as incentives, and customer complaints and customer loyalty are two outcomes of overall customer satisfaction.

After the conceptual framework was established, a lot of research focused on measuring customer satisfaction by quantitative methods, including the analytic hierarchy process (AHP), grey relational analysis and the structural equation model (SEM). In regard to AHP, Li et al. (*7*) proposed a customer satisfaction evaluation method for customized product development. They determined criteria weights through integrating entropy weight and the AHP method. Lan et al. (*8*) introduced a model using the triangular fuzzy concept to determine the weight of each index. A case study from two large express companies was used to demonstrate the feasibility and practicality of the proposed model for examining customer satisfaction. As for the method of grey relational analysis, Perng et al. (*9*) adopted modified grey relational analysis to establish the priority relations of product categories for shopping purposes and satisfaction. Martinez et al. (*10*) used the grey relational analysis to gauge the levels of service satisfaction of the banks and compared the effects of different variables on service satisfaction. However, these two approaches are both devoted to calculating the weights of variables from a subjective perspective. When there are many indicators, optimal values and judgment matrix are difficult to estimate.

As a result, SEM becomes more and more popular because it can increase the validity of a coefficient by reducing measurement error (*11, 12*). As an extension of the traditional regression model, SEM combines path estimation, factor analysis and regression analysis (*13*). Most importantly, SEM can deal with several dependent variables and analyze the relationship between each two factors. As the assessment of customer satisfaction can be complex, with some variables being both dependent and independent, SEM can be used to examine a network of interrelated variables with direct or indirect connections. SEM takes several manifest variables as the indicators of certain latent variables whose influences cannot be measured directly (*14, 15*). So the model consists of two parts: the measurement model which explains the relationship between manifest variables and latent variables, and the structural model which presents the relationship between latent variables.

In the field of transportation, close attention has been paid to public transit passengers' satisfaction in recent years. The user satisfaction of a public transport system is defined as the overall level of attainment of a customer's expectation, measured as the percentage of the expectations actually fulfilled (*16*). Realizing what elements and to what degree they influence people's feeling of taking public transit would help improve the systems, thus attracting more passengers (*17*). The most commonly used method in the literature is SEM (*18–25*). For instance, DeFrancisco et al. (*23*) tried to find out the association between passengers' current travel attitude, demographics and their attitudes toward the current system. Shaaban et al. (*24*) presented a SEM to show that three key factors—namely, service at bus stops, bus service, and service of bus drivers—were strongly correlated to the mode of choice. However, the parameter estimation method of SEM in previous papers is mainly based on maximum likelihood (ML). ML estimation relies on the hypothesis that observed data obey multivariate normal distribution which is not easy to meet in reality (*26*). Moreover, it would lead to unreliable results when observed variables are not continuous or the sample is not big enough (*27*). Partial least square (PLS) can solve ML estimation drawbacks and allow ordinal discrete variables to be put into the model. Also, the sample size can be smaller than 500 using PLS estimation (*28*). Shen et al. (*25*) first compared PLS and ML estimation in the case study of Suzhou urban rail transit and found that the PLS model provided more reasonable parameters. Nonetheless, Shen still followed Fornell's original framework. As a result, passengers' demographics and travel characteristics were not included in Shen's model. Meanwhile, most studies based on the SEM method often treated passengers as a whole group and did not have an insight into the influence of passengers with different usage patterns.

Given this background, using the PLS estimation method this paper aims to measure the degree of satisfaction of cyclists. Cyclists' current usage characteristics and demographics are also included in the model, thus depicting a more comprehensive structure compared to previous studies. Additionally, the models of commuters and non-commuters are calibrated separately so that the perception difference between these two groups of people can be determined.

Methodology

Data Preparation

Before the formal survey, a preliminary list of indicators were developed which were meant to be put in the questionnaire according to respondents' ideas during pre-survey. Then twenty experienced managers from traffic departments, free-floating bike-sharing enterprises and residential communities were interviews. Managers graded each of the indicators (5 class marking system) according to necessity and importance from their point of view. Some indicators were removed and some were combined based on the managers' opinions. A questionnaire was conducted from May 15, 2017 to June 30, 2017. There were 32 survey sites, which comprised 12 main rail transit stations, 8 large residential areas, 7 large companies and 5 university campuses. Therefore, the respondents were restricted to university students, enterprise staff, self-employed entrepreneurs and retirees. Of the 800 questionnaires sent out, 768 were returned. The final available sample size was 724. Travelers who had used free-floating bicycles were defined as respondents and asked to fill in questionnaires. The questionnaire contained four parts: (a) demographics: gender, age, occupation, educational level, income per month; (b) current usage characteristics: average travel time, average travel distance, average parking time and usage frequency per week; (c) respondents' degree of satisfaction with 11 aspects of FFBS, comprising cleanliness of bicycle, convenience of bicycle usage, riding environment, riding safety, accessibility, charging standard, laws and regulations, staff service, parking conditions, information distribution, and government intervention; (d) respondents' satisfaction, loyalty and complaints about FFBS. Most questions were rated on a 5-point scale with 1 denoting "strongly dissatisfied" and 5 denoting "strongly satisfied," except for questions related to individual information which were rated according to the classification presented in Table 1. Taking income as an example, it is a 4-point scale question with 1 denoting "<4,000 RMB" and 4 denoting ">12,000 RMB."

Descriptive Analysis

According to the statistical data obtained from the survey, some traits of FFBS users can be summarized (Table 1). People who are under the age of 40 account for 67% of the users among the respondents. People 40 years of age or older may not choose to use this system because of not having mobile phones or having a physical restriction. Income levels of the respondents showed that 40% earn more than 8,000 RMB when the average monthly income in Shanghai is about 6,500 RMB (*29*). Additionally, most of them (88%) have a bachelor's degree or above. It indicates that people with a good education background tend to have higher acceptance of FFBS. A significant portion of cyclists (39%) use free-floating bicycles more than six times a week, but the average travel distance per time is within 3 km (75%).

When it comes to occupation, 55% of people surveyed are enterprise staff who work fixed hours and need to commute

Table 1. Demographics and Usage Characteristics of Survey Respondents

Attribute	Proportion (%)
Gender	
Male	53
Female	47
Age	
<20	12
20–29	35
30–39	32
40–50	12
>50	9
Education level	
Senior high school or below	12
Undergraduate	45
Master's degree	32
Doctoral degree or above	11
Income per month	
<4,000 RMB	28
4,000–7,999 RMB	32
8,000–11,999 RMB	28
>12,000 RMB	12
Occupation	
University student	21
Enterprise staff	55
Self-employed entrepreneurs	18
Retiree	6
Average travel time using free-floating bicycle	
<5 min	13
5–10 min	62
11–19 min	21
20–30 min	3
>30 min	1
Average travel distance using free-floating bicycle	
<3 km	75
3–5 km	21
6–10 km	3
>10 km	1
Average parking time	
<5 min	81
5–10 min	12
11–15 min	3
>15 min	4
Use frequency per week	
1–2 times	15
3–5 times	36
6–10 times	32
>10 times	17

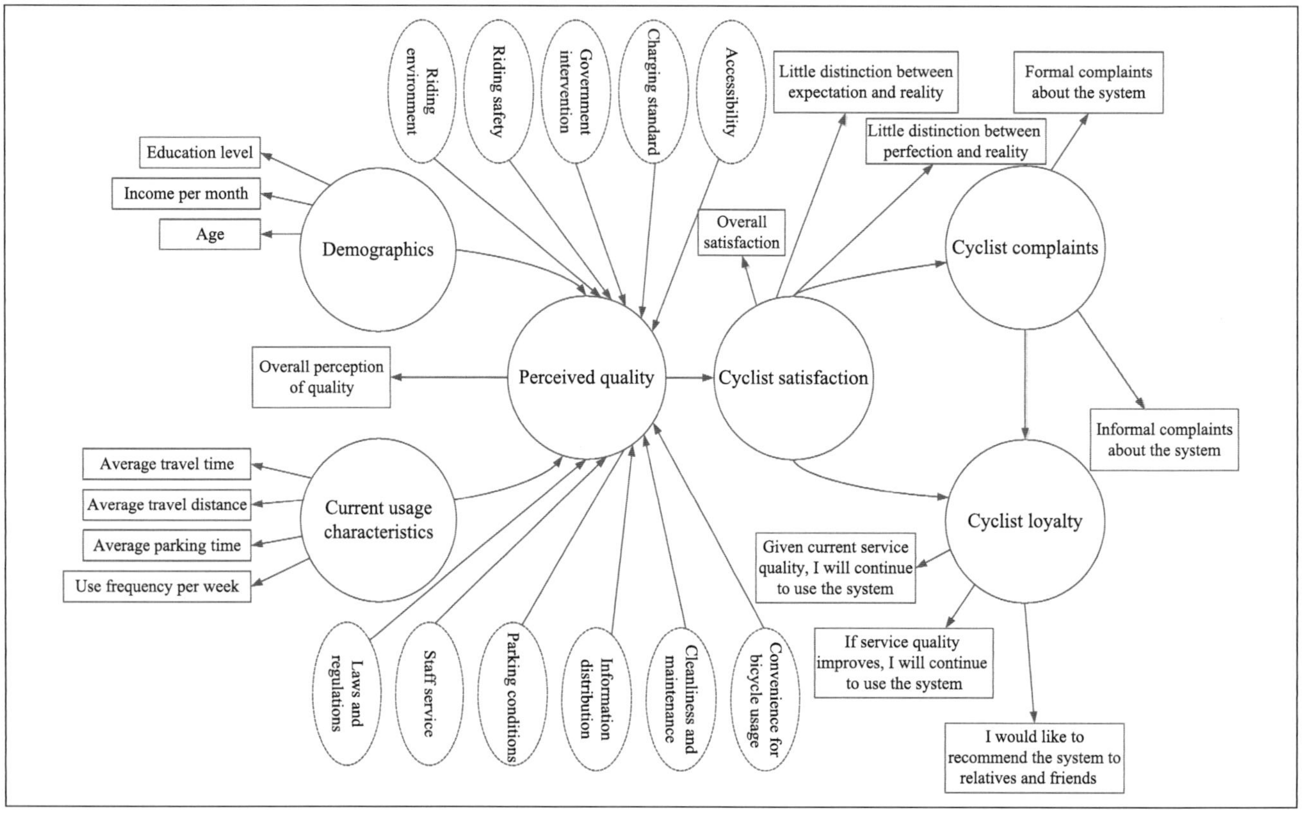

Figure 1. Conceptual framework of the CSI model for FFBS.

during peak hours every work day. The rest of respondents, who are university students (21%) or self-employed entrepreneurs (18%) do not have fixed working hours, so their trip schedules are more flexible. As for retirees, who only account for 6% of respondents, they usually travel for shopping, visiting relatives or some other recreation purpose. As a result, enterprise staff are defined as commuters with the others as non-commuters. Liu et al. (*30*) claimed that commuters and non-commuters have perception differences toward the external environment. In this paper, a comparison between commuters and non-commuters of their perception of service quality and degree of satisfaction with the FFBS is performed.

Model Construction

As shown in Figure 1, the CSI model has three levels. The first level describes the relationship between demographics, current usage characteristics, perceived quality, cyclist satisfaction, loyalty and complaints. Cronin et al. (*31*) found that there is a positive correlation between service quality and satisfaction. At the same time, satisfaction is regarded as the overall affective response to a perceived discrepancy between prior expectations and perceived performance after consumption (*32*). The association between satisfaction, loyalty and complaints is well documented by the literature (*33–35*). Therefore, four hypotheses, H1–H4, are described as follows:

H1: Perceived quality positively influences cyclist satisfaction
H2: Cyclist satisfaction positively influences cyclist loyalty
H3: Cyclist complaints negatively influence cyclist loyalty
H4: Cyclist satisfaction negatively influences cyclist complaints

DeFrancisco et al. (*23*) mentioned that users' demographics such as annual income, age and car ownership affected their attitude toward carpooling. Deutsch et al. (*36*) pointed out that demographics play an important role in understanding people's interaction with their surroundings. Similarly, in this case, cyclists' socioeconomic characteristics and their current usage status can be treated as two elements which influence perceived quality. Therefore, H5 and H6 are presented as follows:

H5: Demographics positively influence perceived quality
H6: Current usage characteristics positively influence perceived quality

The second level of the model is the relationship between perceived quality and 11 latent variables which are subject to it. Based on previous literature, the selection of attributes differed from person to person. For example, Eboli and Mazzulla (*37*) referred to 11 dimensions including cleanliness, comfort, information, safety and so forth. Liou et al. (*38*) included convenience, driver attitude, equipment and reliability when evaluating bus drivers' service quality. In this study, subordinate latent variables were chosen partly based on our literature review. At the same time, the traits of FFBS which are different from other transportation systems were extracted. They are cleanliness and maintenance, convenience of bicycle usage, riding environment, riding safety, accessibility, charging standard, laws and regulations, staff service, parking conditions, information distribution, and government intervention. The hypotheses H7–H17 are presented as follows:

H7: Cleanliness and maintenance positively influence perceived quality
H8: Convenience of bicycle usage positively influences quality
H9: Riding environment positively influences perceived quality
H10: Riding safety positively influences perceived quality
H11: Accessibility positively influences perceived quality
H12: Charging standard positively influences perceived quality
H13: Laws and regulations positively influence perceived quality
H14: Staff service positively influences perceived quality
H15: Parking conditions positively influence perceived quality
H16: Information distribution positively influences perceived quality
H17: Government intervention positively influences perceived quality

Lastly, the association between the latent variables and their manifest variables are defined as the third level of the model. There are 50 manifest variables considered in this research. These variables are attached to different latent variables. For example, formal complaints about the system (P_1) and informal complains about the system (P_2) both belong to the latent variable cyclist complaints (CCO). Others abbreviations and their corresponding manifest variables are shown in Table 4.

The structural part of the model can be calculated as follows:

$$\begin{bmatrix} CLO \\ CCO \\ CSA \\ PER \end{bmatrix} = \begin{bmatrix} 0 & \beta_{12} & \beta_{13} & 0 \\ 0 & 0 & \beta_{23} & 0 \\ 0 & 0 & 0 & \beta_{34} \\ 0 & 0 & 0 & 0 \end{bmatrix} \begin{bmatrix} CLO \\ CCO \\ CSA \\ PER \end{bmatrix} + \begin{bmatrix} 0 & 0 & \cdots & 0 & 0 \\ 0 & 0 & \cdots & 0 & 0 \\ 0 & 0 & \cdots & 0 & 0 \\ \gamma_{41} & \gamma_{42} & \cdots & \gamma_{411} & \gamma_{412} \end{bmatrix} * \begin{bmatrix} DEM \\ CUR \\ \cdots \\ INF \\ GOV \end{bmatrix} + \begin{bmatrix} \zeta_{11} \\ \zeta_2 \\ \zeta_3 \\ \zeta_4 \end{bmatrix} \quad (1)$$

where

β_{ij} = path coefficients of endogenous latent variables;
γ_{ij} = path coefficients of exogenous latent variables;
ζ_i = residuals of endogenous latent variables.

The measurement part of the model can be estimated as follows:

$$\begin{bmatrix} A_1 \\ \cdots \\ D_1 \\ \cdots \\ J_1 \\ \cdots \\ K_1 \\ L_1 \\ \cdots \\ P_1 \end{bmatrix} = \left[\begin{bmatrix} 0 & 00 & 0 & \alpha_{A1} & 0 & 0 & 0 & 0 \\ \cdots & \cdots & \cdots & \cdots & \cdots & \cdots & \cdots & \cdots \\ 0 & 0 & 0 & 0 & 0 & \alpha_{D1} & 0 & 0 \\ \cdots & \cdots & \cdots & \cdots & \cdots & \cdots & \cdots & \cdots \\ 0 & 0 & 0 & 0 & 0 & 0 & \alpha_{J1} & 0 \\ \cdots & \cdots & \cdots & \cdots & \cdots & \cdots & \cdots & \cdots \\ 0 & 0 & 0 & 0 & 0 & 0 & 0 & \alpha_{K1} \\ \alpha_{L1} & 00 & 0 & 0 & 0 & 0 & 0 & 0 \\ \cdots & \cdots & \cdots & \cdots & \cdots & \cdots & \cdots & \cdots \\ 0 & 0 & \alpha_{P1} & 0 & 0 & 0 & 0 & 0 \end{bmatrix} \begin{bmatrix} CLO \\ \cdots \\ PER \\ CLE \\ \cdots \\ RIS \\ INF \\ GOV \end{bmatrix} + \begin{bmatrix} \varepsilon_{A1} \\ \cdots \\ \varepsilon_{D1} \\ \cdots \\ \varepsilon_{J1} \\ \cdots \\ \varepsilon_{K1} \\ \varepsilon_{L1} \\ \cdots \\ \varepsilon_{P1} \end{bmatrix}\right] \quad (2)$$

where

α_{yi} = loading coefficients of all 50 manifest variables;
ε_{yi} = measuring errors.

Validity and Reliability Test

The calibration of the CSI model is based on the software SmartPLS3.0. Fit devices are adopted to examine the appropriateness of the model. Cronbach's Alpha (CA) is a widely used index to measure how well the manifest variables explain their corresponding latent variables. Generally, when CA is greater than 0.35, the model is considered to be acceptable (*39*). Composite reliability (CR) and average variance extracted (AVE) are also two vital indexes. CR is required to be greater than 0.6 and AVE should be over 0.5 (*27, 40*).

The initial estimation results show that CA, CR and AVE for cyclist complaints are far below the criteria which means that the two manifest variables, "Formal complaints about the system" and "Informal complains about the system," cannot clearly explain cyclist complaints. This may be because respondents have difficulty in understanding the discrepancy between formal and informal complaints. Therefore, the unreliable construct, "cyclist complaints," is excluded from the model and the final results of model fit are shown in Table 2. Fit devices of all

Table 2. Fit Devices of the Model

Group name	Commuters			Non-commuters		
Fit device	AVE	CR	CA	AVE	CR	CA
Cleanliness and maintenance	0.560	0.772	0.613	0.584	0.848	0.762
Convenience of bicycle usage	0.595	0.854	0.774	0.655	0.883	0.829
Riding environment	0.558	0.796	0.671	0.693	0.919	0.889
Riding safety	0.670	0.858	0.755	0.766	0.907	0.846
Accessibility	0.651	0.848	0.731	0.741	0.853	0.660
Charging standard	0.594	0.805	0.634	0.690	0.869	0.690
Laws and regulations	0.760	0.905	0.841	0.778	0.913	0.857
Staff service	0.721	0.886	0.807	0.733	0.892	0.818
Parking conditions	0.844	0.915	0.818	0.827	0.905	0.791
Information distribution	0.784	0.879	0.726	0.837	0.911	0.807
Government intervention	0.816	0.775	0.775	0.527	0.710	0.558
Cyclist loyalty	0.737	0.894	0.822	0.626	0.832	0.704
Cyclist satisfaction	0.629	0.835	0.705	0.696	0.873	0.782
Current usage situation	0.578	0.622	0.441	0.579	0.635	0.531
Demographics	0.651	0.691	0.516	0.673	0.608	0.426
Perceived quality	1.000	1.000	1.000	1.000	1.000	1.000

the latent variables meet the test criterion. What's more, p-value of each path is lower than 0.05 (shown in Table 3), thus proving the validation of the model.

Model Results and IPA Analysis

According to Table 3, all of the hypotheses proposed in this part of the model construction are proved to be reasonable, although the correlation between demographics and perceived quality is negative, which is contrary to the original hypothesis. In fact, it shows that cyclists with increased age, higher income and higher education have stricter FFBS quality requirements. This phenomenon appears more obvious among commuters, as the path coefficient (−0.018) is more than twice as much as that of non-commuters (−0.007). As for current usage characteristics, shortening the average parking time will improve cyclists' perception of service quality, with −0.485 and −0.705 of loading coefficients for commuters and non-commuters respectively (presented in Table 4). Non-commuters focus more on the time spent and travel distance of bicycle riding. Those who travel longer distances have a better experience because they may acquire a better understanding of the system. However, among commuters, frequency of use has the largest magnitude with a coefficient of 0.816. The reason lies in that, according to the survey, commuters use FFBS more frequently, particularly on weekdays, so their attitude towards the system is more sensitive to the number of times they travel by free-floating bicycle.

To gain a deep understanding of how the 11 dimensions (cleanliness and maintenance, convenience for bicycle usage, riding environment, etc.) interact with perceived quality, a term called satisfaction index (SI) is proposed. SI is a standard for satisfaction evaluation using the centesimal system which can be calculated as follows:

$$SI = \frac{E[Z]-Min[Z]}{Max[Z]-Min[Z]} \times 100 \tag{3}$$

where

Z = latent variable affiliated to perceived quality;
$E[\,.\,]$ = expected value of the variable;
$Max[\,.\,]$ = maximum value of the variable;
$Min[\,.\,]$ = minimum value of the variable.

$$Min[Z] = \sum_{i=1}^{n} w_i Min[x_i] \tag{4}$$

$$Max[Z] = \sum_{i=1}^{n} w_i Max[x_i] \tag{5}$$

where

x_i = manifest variables of the latent variable;
w_i = weights;
n = number of manifest variables.

In this study, formulas (3)–(5) can be simplified given the 5-point scale:

$$SI = \frac{\sum_{i=1}^{n} w_i \overline{x_i} - \sum_{i=1}^{n} w_i}{4\sum_{i=1}^{n} w_i} \times 100 \tag{6}$$

The results of the 11 perspectives of perceived quality measured in this way are demonstrated in Table 5. A score less than 60 means cyclists are dissatisfied with this aspect of the FFBS. A score between 60 and 75 represents "moderately satisfied" and a score between 75 and 90 means "relatively satisfied." When the score is over 90, it means cyclists are

Table 3. Verification of SEM Hypotheses

Path relation	Path coefficient		P-value		Remark
	For commuters	For Non-commuters	For commuters	For Non-commuters	
Demographics→Perceived quality	−0.018	−0.007	0.031	0.036	Accept
Current usage characteristics→Perceived quality	0.040	0.013	0.037	0.027	Accept
Perceived quality→Cyclist satisfaction	0.468	0.681	0.000	0.000	Accept
Cyclist satisfaction→Cyclist loyalty	0.673	0.482	0.000	0.000	Accept
Cleanliness and maintenance→Perceived quality	0.014	0.127	0.040	0.000	Accept
Convenience for bicycle usage→Perceived quality	0.088	0.093	0.017	0.000	Accept
Riding environment→Perceived quality	0.386	0.236	0.000	0.000	Accept
Riding safety→Perceived quality	0.104	0.130	0.013	0.000	Accept
Accessibility→Perceived quality	0.109	0.105	0.036	0.000	Accept
Charging standard→Perceived quality	0.123	0.086	0.030	0.000	Accept
Laws and regulations→Perceived quality	0.237	0.126	0.002	0.000	Accept
Staff service→Perceived quality	0.175	0.137	0.009	0.000	Accept
Parking conditions→Perceived quality	0.043	0.094	0.019	0.001	Accept
Information distribution→Perceived quality	0.016	0.047	0.048	0.017	Accept
Government intervention→Perceived quality	0.133	0.195	0.028	0.016	Accept

very satisfied with this aspect. Commuters' and non-commuters' degrees of satisfaction are basically consistent with each other. In particular, parking conditions and staff service perform badly, with SI lower than 60, which means the current conditions of these two service dimensions need to be improved. The remaining nine dimensions perform relatively well. However, the scores are all below 90. Charging standard has the highest SI among commuters, whereas convenience of bicycle usage is the most satisfying for people do not commute on weekdays. In general, non-commuters think more highly of FFBS than commuters when comparing the average values of SI across the 11 dimensions. One of the probable causes could be that most of their trips are not as urgent as commuters'. As a result, they seem to be more tolerant of the shortcomings of the system.

Importance-performance analysis (IPA) is a useful quantitative approach to identifying the key points of improvement for the FFBS. The performance is the SI of the 11 dimensions of perceived quality and the importance is presented by path coefficients. The results of IPA are illustrated in Figure 2, with the *x*-axis denoting the importance and the *y*-axis the performance. The four quadrants are divided based on the mean value of SI as well as the mean value of the path coefficients of the 11 service aspects. They are marked as I, II, III, and IV, which mean "Keep up the good work," "Concentrate here," "Lower priority," and "Possible overkill" respectively (*41*). Obviously, indicators in Quadrant II need the most attention because they perform badly at present but have the greatest impact on the system.

From the perspective of commuters (Figure 2*a*), riding environment and laws and regulations are in Quadrant II which means commuters think the problems of these two indicators must be solved immediately, or they will became the major stumbling block of the FFBS. In Quadrant III, there are three indicators including parking conditions, staff service and riding safety, which indicates that they perform badly but have a lower priority for improved. The remaining indicators are all in Quadrant IV. That is to say, they just need to maintain the present state. However, the non-commuters' attitude (Figure 2*b*) is not the same as the commuters'. From their point of view, three dimensions including government intervention, riding safety and staff service should be added to Quadrant II, because among the non-commuters surveyed, university students and retirees occupied a fairly large proportion. University students were the initial users of the FFBS in Shanghai, as FFBS companies such as OFO first put free-floating bicycles on school campuses. So they may have more requirements of the FFBS after having used the system for a while. With regard to the retirees, they prefer to have better service because they may have more difficulty using the app compared to young people. Meanwhile, they give priority to riding safety because of their physical limitations.

Conclusion and Future Work

The objective of this research was to assess the potential of the FFBS and how to increase its usage and effectiveness. To achieve these goals, a questionnaire was conducted to answer

Table 4. Loading Coefficients of Manifest Variables

Latent variables	Manifest variables	Loading coefficients	
		For commuters	For non-commuters
Cleanliness and maintenance (CLE)	Cleanliness of bicycle (A1)	0.665	0.726
	Design of bicycle (A2)	0.703	0.795
	Brake sensitivity (A3)	0.617	0.787
	Bicycle integrity (A4)	0.724	0.745
Convenience of bicycle usage (CON)	Time spent renting and returning (B1)	0.640	0.866
	Operation of smart lock (B2)	0.804	0.758
	Process of code scanning (B3)	0.829	0.825
	Process of payment (B4)	0.799	0.785
Riding environment (RIE)	Flatness of bicycle lane (C1)	0.775	0.804
	Width of bicycle lane (C2)	0.496	0.859
	Shade facilities (C3)	0.372	0.830
	Connectivity of bicycle lane (C4)	0.762	0.829
	Traffic density in bicycle lane (C5)	0.851	0.841
Riding safety (RIS)	Separation facilities (D1)	0.756	0.821
	Other non-motorized vehicle users' manners (D2)	0.846	0.902
	Motorized vehicle users' manners (D3)	0.867	0.899
Accessibility (ACC)	Quantity of bicycles (E1)	0.863	0.752
	Positioning accuracy (E2)	0.827	0.852
	Acceptability of walking distance (E3)	0.724	0.830
Charging standard (CHA)	Reasonableness of the deposit (F1)	0.883	0.839
	Reasonableness of the rent per time (F2)	0.490	0.787
	Degree of preferential benefit (F3)	0.873	0.864
Laws and regulations (LAW)	Claim mechanism (G1)	0.846	0.901
	Punishment mechanism (G2)	0.892	0.870
	Integrity of regulations (G3)	0.876	0.875
Staff service (STA)	Efficiency of dispatch (H1)	0.823	0.822
	Efficiency of removing broken bicycles (H2)	0.868	0.875
	Response to users' enquiries (H3)	0.856	0.871
Parking conditions (PAR)	Adequacy of parking areas (I1)	0.896	0.910
	Parking order (I2)	0.941	0.909
Information distribution (INF)	Accessibility of the latest information (J1)	0.897	0.894
	Publicity of launch plan (J2)	0.874	0.935
Government intervention (GOV)	Constraint force over bicycle company law (K1)	0.901	0.872
	Punctuality of management policy notification (K2)	0.905	0.598
Cyclist loyalty (CLO)	Given current service quality, I will continue to use the system (L1)	0.843	0.871
	If service quality improves, I will continue to use the system (L2)	0.845	0.672
	I would like to recommend the system to relatives and friends (L3)	0.887	0.817
Cyclist satisfaction (CSA)	Overall satisfaction (M1)	0.795	0.844
	Little distinction between expectation and reality (M2)	0.703	0.812
	Little distinction between perfection and reality (M3)	0.872	0.846
Current usage situation (CUR)	Average travel time (N1)	0.317	0.748
	Average travel distance (N2)	0.458	0.676
	Average parking time (N3)	−0.485	−0.705
	Use frequency per week (N4)	0.816	0.352
Demographics (DEM)	Education level (O1)	0.388	0.961
	Income per month (O2)	0.890	0.636
	Age (O3)	0.642	0.302
Perceived quality	Overall perception of quality (R1)	1.000	1.000

Table 5. Satisfaction Indexes of Latent Variables

	Commuters		Non-commuters	
Latent variable	SI	Degree of satisfaction	SI	Degree of satisfaction
Cleanliness and maintenance	67.38	Moderately satisfied	71.32	Moderately satisfied
Convenience of bicycle usage	75.96	Relatively satisfied	81.69	Relatively satisfied
Riding environment	62.07	Moderately satisfied	62.86	Moderately satisfied
Riding safety	64.19	Moderately satisfied	63.42	Moderately satisfied
Accessibility	68.88	Moderately satisfied	72.17	Moderately satisfied
Charging standard	76.88	Relatively satisfied	78.64	Relatively satisfied
Laws and regulations	62.14	Moderately satisfied	61.30	Moderately satisfied
Staff service	55.67	Dissatisfied	59.77	Dissatisfied
Parking conditions	50.52	Dissatisfied	58.43	Dissatisfied
Information distribution	71.85	Moderately satisfied	68.36	Moderately satisfied
Government intervention	67.28	Moderately satisfied	66.01	Moderately satisfied

three main questions: "Who is using the FFBS?" "What are the demographics and travel characteristics of FFBS users?" and most importantly, "What factors influence their degree of satisfaction of the system?"

According to the descriptive analysis, most users are aged between 20 and 40, with relatively high education levels. The majority of them earn less than 12,000 RMB monthly, which represents a middle income level in Shanghai. The travel time is often 20 minutes or less and the average travel distance is mostly below 5 km. SEM was used as a tool to analyze how the elements inside the system interacted with each other. The results show that demographics have a negative relationship and users' current travel characteristics have a positive relationship with perceived quality. The direct impact of cyclist satisfaction on cyclist loyalty is very large. IPA analysis reveals that five dimensions of the FFBS must be improved urgently. They are riding environment, riding safety, government intervention, laws and regulations, and staff service.

Based on findings from the research, some important policy implications can be determined to help optimize the system. First of all, to optimize the current riding environment and safety situation, Tironi (*42*) once suggested that quality of bicycle facilities affected the amount of people who traveled by bikes. Using the Shanghai statistics, the mean width of non-motorized lanes is 3.1 m. However, there is a big difference between arterial roads (3.5 m) and branch roads (2.45 m). Moreover, the width of non-motorized lanes demonstrates a decrease from the urban area to the outskirts. Therefore, there is a need to provide a higher level of service on the branch roads by widening the cycle lanes, especially for those in the suburbs. Connectivity of bicycle lanes should also be strengthened.

At present, government enforcement related to the FFBS is relatively weak. How to balance the relationships among governors, cyclists, and FFBS enterprises is a key issue to be solved. For example, the responsibilities and obligations of FFBS companies must be clarified in writing. The number of FFBS companies and the number of bicycles they have should be controlled, within reason. For this purpose, governors have been working an information platform, which allows the real-time data sharing. Data retrieved from GPS devices installed in free-floating bicycles will be sent to the platform. Managers will obtain an understanding of bicycle routes, the number of bicycles in service, real-time locations and bicycle fault information. Ranking of all companies is assessed monthly according to their operational situation and problem-solving efficiency. In addition, cyclists' behaviors should also be supervised via the establishment of a credit mechanism so that violations of regulations can be punished by reducing credit scores or fining the cyclists.

Meanwhile, electronic fences are being built in several pilot areas. When cyclists park the free-floating bicycles in places where parking is forbidden, the built-in smart device will identify the mistake and remind users. Users will continue to be billed until they park bicycles in right the place.

Although the study has made some discoveries, several limitations still exist. First, "cyclist complaints" was excluded from the model initially because of the deficiency of the questionnaire design, but it is an important aspect which reflects cyclists' dissatisfaction with services. Second, respondents who were surveyed this time only included four occupations, which may lead to deviation in model results. In that case, further investigation and better survey design are needed to improve the CSI model in the future.

Acknowledgments

This study is supported by the Humanities & Social Sciences Project of the Chinese Ministry of Education (17YJC630179), Key Project of National Natural Science Foundation(71734004) and the Fundamental Research Funds for the Central Universities (20160629 and wx0160020170889).

References

1. Shaheen, S. A., S. Guzman, and H. Zhang. Bikesharing in Europe, the Americas, and Asia: Past, Present, and Future. *Transportation Research Record: Journal of the Transportation Research Board*, 2011. 2143: 159–167.

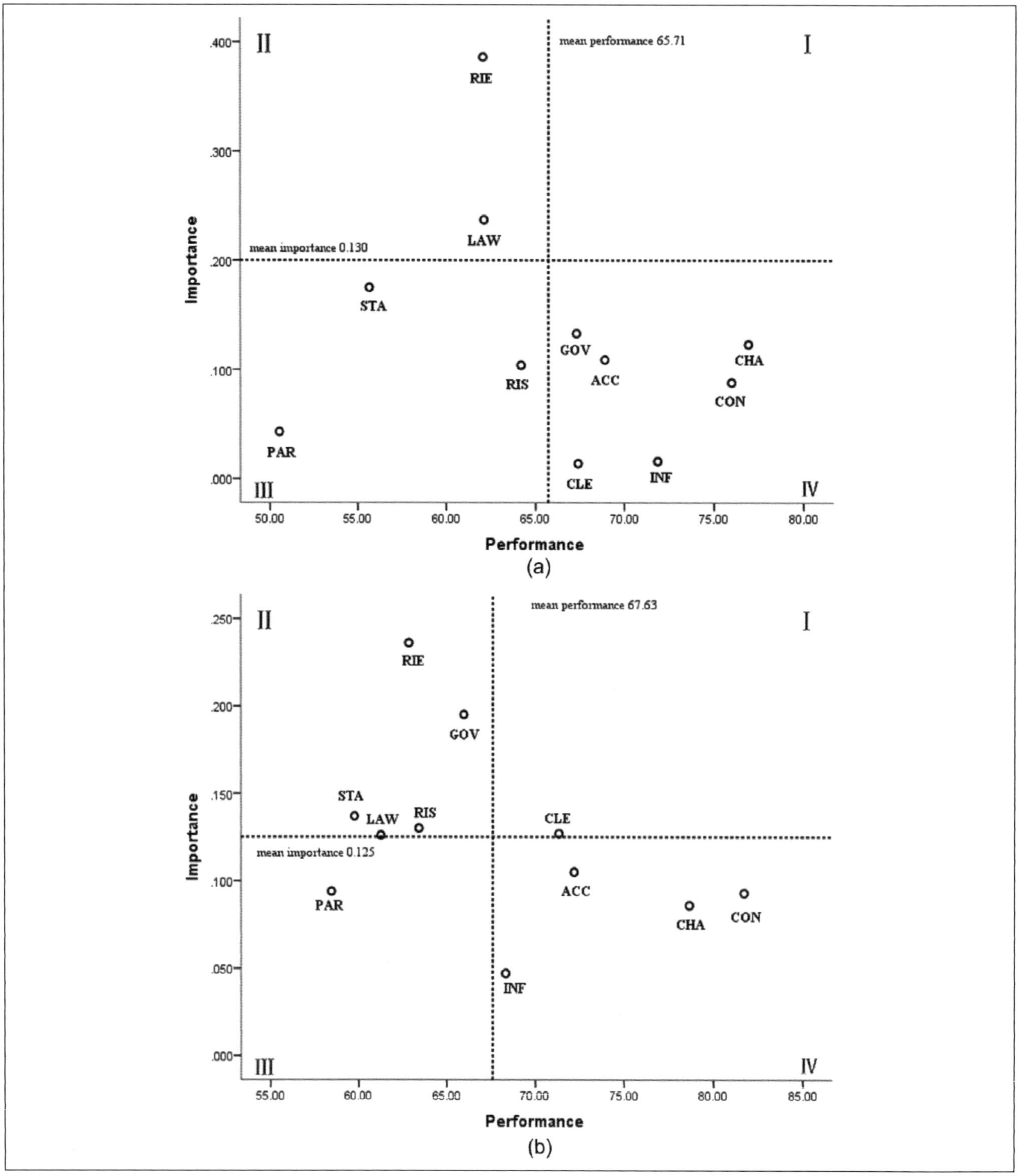

Figure 2. (*a*) IPA matrix result for commuters, and (*b*) IPA matrix result for non-commuters.

2. Daedozo, R. N. An Experimental Study of Consumer Effort, Expectation, and Satisfaction. *Journal of Marketing Research*, Vol. 2, 1965, pp. 244–249.
3. Cronin, J. J., and S. A. Taylor. Measuring Service Quality: A reexamination and Extension. *Journal of Marketing*, Vol. 56, 1992, pp. 55–68.
4. Boulding, W., A. Kalra, R. Staelin, and V. A. Zeithaml. A Dynamic Process Model of Service Quality: From Expectations to Behavioral Intentions. *Journal of Marketing Research*, Vol. 30, No. 3, 1993, pp. 7–27.
5. Eugene, W. A., F. Claes, F., and D. R. Lehmann. Customer Satisfaction, Market Share, and Profitability: Findings from

Sweden. *Journal of Marketing*, Vol. 58, No. 3, 1994, pp. 53–66.

6. Fornell, C., M. D. Johnson, E. W. Anderson, J. Cha, and B. E. Bryant. The American Customer Satisfaction Index: Nature, Purpose, and Findings. *Journal of Marketing*, Vol. 60, No. 4, 1996, pp. 7–18.
7. Li, L., F. Liu, and C. Li. Customer Satisfaction Evaluation Method for Customized Product Development Using Entropy Weight and Analytic Hierarchy Process. *Computers & Industrial Engineering*, Vol. 77, 2014, pp. 80–87.
8. Lan, S., H. Zhang, R. Y. Zhong, and G. Q. Huang. A Customer Satisfaction Sevaluation Model for Logistics Services Using Fuzzy Analytic Hierarchy Process. *Industrial Management & Data Systems*, Vol. 116, No. 5, 2016, pp. 1024–1042.
9. Perng, S. W., C. C. Chow, and W. C. Liao. Analysis of Shopping Preference and Satisfaction with Airport Retailing Products. *Journal of Air Transport Managemen*t, Vol. 16, No. 5, 2010, pp. 279–283.
10. Martínez, L., Silva, J. D. A. E., and J. Viegas. Assessment of Residential Location Satisfaction in the Lisbon Metropolitan Area. *Headache*, Vol. 50, No. 6, 2010, pp. 998–1004.
11. Dion, P. A. Interpreting Structural Equation Modeling Results: A Reply to Martin and Cullen. *Journal of Business Ethics*, Vol. 83, No. 3, 2008, pp. 365–368.
12. Wright, S. Correlation and Causation. *Journal of Agricultural Research*, Vol. 20, 1921, pp. 557–585.
13. Stuart, K., M. Mednick, and J. Bockman. Structural Equation Model of Customer Satisfaction for the New York City Subway System. *Transportation Research Record: Journal of the Transportation Research Board*, 2000. 1735: 133–137.
14. Hoyle, R. H. *Structural Equation Modeling: Concepts, Issues, and Applications*. SAGE, Thousand Oaks, London, 1995.
15. Dion, P. A. Interpreting Structural Equation Modeling Results: A Reply to Martin and Cullen. *Journal of Business Ethics*, Vol. 83, No. 3, 2008, pp. 365–368.
16.Tyrinopoulos, Y., and C. Antoniou. Public Transit User Satisfaction: Variability and Policy Implications. *Transport Policy*, Vol. 15, No. 4, 2008, pp. 260–272.
17. Oña, J. D., R. D. Oña, L. Eboli, and G. Mazzulla. Perceived Service Quality in Bus Transit Service: A Structural Equation Approach. *Transport Policy*, Vol. 29, No. 3, 2013, pp. 219–226.
18. Karlaftis, M., J. Golias, and E. Papadimitriou. Transit Quality as an Integrated Traffic Management Strategy: Measuring Perceived Service. *Journal of Public Transportation*, Vol. 4, No. 1, 2001, pp. 27–44.
19Eboli, L., and G. Mazzulla. Service Quality Attributes Affecting Customer Satisfaction for Bus Transit. *Journal of Public Transportation*, Vol. 10, No. 3, 2007, pp. 21–34.
20. Chen, C. F. Investigating Structural Relationships between Service Quality, Perceived Value, Satisfaction, and Behavioral Intentions for Air Passengers: Evidence from Taiwan. *Transportation Research Part A*, Vol. 42, No. 4, 2008, pp. 709–717.
21. Githui, J. N., T. Okamura, and F. Nakamura. The Structure of Users' Satisfaction on Urban Public Transport Service in Developing Country: The Case of Nairobi. *Journal of the Eastern Asia Society for Transportation Studies*, Vol. 8, 2010, pp. 1288–1300.
22. Irfan, S. M., D. M. H. Kee, and S. Shahbaz. Service Quality and Rail Transport in Pakistan: A Passenger Perspective. *World Applied Sciences Journal*, Vol. 18, No. 3, 2013, pp. 361–369.
23. DeFrancisco, J., R. Harb, and E. Radwan. *Evaluation of a Carpooling Program in a University Setting Using a Stated Preference Survey*. Presented at 93rd Annual Meeting of the Transportation Research Board, Washington, D.C., 2014.
24.Shaaban, K., and I. Kim. The Influence of Bus Service Satisfaction on University Students' Mode Choice. *Journal of Advanced Transportation*, Vol. 50, No. 6, 2016, pp. 935–948.
25.Shen, W., W. Xiao, and X. Wang. Passenger Satisfaction Evaluation Model for Urban Rail Transit: A Structural Equation Modeling Based on Partial Least Squares. *Transport Policy*, Vol. 46, 2016, pp. 20–31.
26. Chin, W. W. Partial Least Squares is to LISREL as Principal Components Analysis is to Common Factor Analysis. *Technology Studies*, Vol. 2, 1995, pp. 315–319.
27. Fornell, C. A Comparative Analysis of Two Structural Equation Models: Lisrel and PLS Applied to Market Data. *Minnesota Medicine*, Vol. 69, No. 1, 1981, pp. 35–38.
28. Hu, L. T., P. M. Bentler, and Y. Kano. Can Test Statistics in Covariance Structure Analysis be Trusted? *Psychological Bulletin*, Vol. 112, No. 2, 1992, pp. 351–362.
29.Shanghai Institute of City and Transport Planning. Main Achievements of the Fifth Shanghai Traffic Survey. *Traffic & Transportation*, Vol. 31, No. 6, 2015, pp. 15–18.
30. Liu, C., Y. O. Susilo, and A. Karlström. Investigating The Impacts of Weather Variability on Individual's Daily Activity–Travel Patterns: A Comparison between Commuters and Non-Commuters in Sweden. *Transportation Research Part A*, Vol. 82, 2015, pp. 47–64.
31. Cronin, J. J., Jr., M. K. Brady, and G. T. M. Hult. Assessing the Effects of Quality, Value, and Customer Satisfaction on Consumer Behavioral Intentions in Service Environments. *Journal of Retailing*, Vol. 76, No. 2, 2000, pp. 193–218.
32. Stoltman, J., and R. Hampton. A Model of the Antecedents and Consequences of Cognitive Response to Puffery in Advertisements. *AMA Winter Educators' Conference Proceedings*, Phoenix, Ariz., 1985. published by Chicago, III.: American Marketing Association, pp. 42–50.
33. Wen, C. H., L. W. Lan, and H. L. Cheng. Structural equation modeling to determine passenger loyalty toward intercity bus services. *Transportation Research Record: Journal of the Transportation Research Board*, 2005. 1927: 249–255.
34. Chou, C. C., L. J. Liu, S. F. Huang, J. M. Yih, and T. C. Han. An Evaluation of Airline Service Quality Using the Fuzzy Weighted SERVQUAL Method. *Applied Soft Computing*, Vol. 11, No. 2, 2011, pp. 2117–2128.
35. Chou, J. S., and C. Kim. A Structural Equation Analysis of the QSL Relationship with Passenger Riding Experience on High Speed Rail: An Empirical Study of Taiwan and Korea. *Expert Systems with Applications*, Vol. 36, No. 3, 2009, pp. 6945–6955.
36. Deutsch, K., S. Y. Yoon, and K. Goulias. Modeling Travel Behavior and Sense of Place Using a Structural Equation Model. *Journal of Transport Geography*, Vol. 28, No. 3, 2013, pp. 155–163.
37. Eboli, L., and G. Mazzulla. A Methodology for Evaluating Transit Service Quality Based on Subjective and Objective Measures from the Passenger's Point of View. *Transport Policy*, Vol. 18, No. 1, 2011, pp. 172–181.
38. Liou, J. J. H., C. C. Hsu, and Y. S. Chen. Improving Transportation Service Quality Based on Information Fusion. *Transportation Research Part A*, Vol. 67, No. 67, 2014, pp. 225–239.

39. Cronbach, L. J. Coefficient Alpha and the Internal Structure of Tests. *Psychometrika*, Vol. 16, No. 3, 1951, pp. 297–334.
40. Bagozzi, R. P., and Y. Yi. On the Evaluation of Structural Equation Models. *Journal of the Academy of Marketing Science*, Vol. 16, No. 1, 1988, pp. 74–94.
41. Figler, S. A., P. S. Iraj, E. W. Welch, and N. Yavuz. Customer Loyalty and Chicago, Illinois, Transit Authority Buses. *Transportation Research Record: Journal of the Transportation Research Board*, 2015. 2216: 148–156.
42. Tironi, M. Ethics of Care of Urban Resources: Maintenance and Repair on a Public Bicycle System. *ARQ*, Vol. 89, 2015, pp. 76–89.

The Standing Committee on Transportation in the Developing Countries (ABE90) peer-reviewed this paper (18-05091).

Article

TRR
JOURNAL OF THE TRANSPORTATION RESEARCH BOARD

Transportation Research Record
2018, Vol. 2672(31) 33–40

DOI: 10.1177/0361198118757996
journals.sagepub.com/home/trr
SAGE

Study on the Invasion Behavior of E-Bikes with Motor Vehicle Traffic at a Signalized Intersection

Da Yang[1,2], Xiaoxia Zhou[1], Danhong Wu[1], and Sijing Liu[1]

Abstract

In many developing countries like China, many queuing electric bikes (e-bikes) passing an intersection simultaneously greatly reduces the capacity of the intersection for motor vehicles, by invading the passing area of motor vehicles. To study the invasion effect of e-bikes on the traffic flow of motor vehicles at an urban signalized intersection, this paper proposes a social force model for the heterogeneous traffic flow of motor vehicles and e-bikes. The proposed model is calibrated and validated using real data collected in Chengdu, China. The validation results show that the proposed model can replicate the heterogeneous traffic flow with low errors. Simulations based on the proposed model are conducted to investigate what strategies can reduce the invasion of e-bikes in normal motor vehicle traffic. The results show that when the number of queuing e-bikes before the stop line is more than 20, the two strategies can be applied: the stop-line-ahead strategy and the green-signal-ahead strategy. The study suggests that the 2–4 s of green signal ahead or 3–5 m of stop line ahead for non-motor vehicles can significantly reduce the interference of e-bikes on motor vehicle traffic. In addition, the combination of the two strategies can also obtain the same effect but with smaller change to the original intersection design.

The electric bike (e-bike) is a widely used mode of travel in some developing countries like China. E-bikes can provide a relatively convenient and inexpensive traffic tool for commuters (*1*). However, e-bikes also bring many traffic problems. At a signalized intersection, when many e-bikes pass through the intersection simultaneously they will invade the passing space of motor vehicles and cause much traffic conflict between e-bikes and motor vehicles. Such a phenomenon can heavily reduce the intersection capacity, and finding ways to solve the problem is important to improve the efficiency of an intersection.

As e-bikes have the properties of small size, light weight, and high starting speeds, they have greater flexibility than motor vehicles and normal bicycles at intersections. Once the signal turns to green, e-bikes can accelerate to a higher speed than motor vehicles in a short period, and then quickly occupy the passing space of motor vehicles, producing a heterogeneous traffic flow of motor vehicles and e-bikes, as shown in Figure 1. Therefore, establishing a model of heterogeneous traffic flow of motor vehicles and e-bikes, and studying the impact of straight-moving e-bikes on motor vehicle traffic flow, are critical to solving this problem. The results of the study will provide appropriate theoretical support for intersection design.

To date, studies on the mixed traffic flow of e-bikes and motor vehicles at intersections are rare. Existing studies on mixed traffic at intersections have mainly focused on the interaction characteristics between straight-moving normal bicycles and left-turn cars or right-turn cars (*2–4*). The existing methods of modeling the mixed traffic flow of bicycles and cars at intersections can be classified into the three categories: longitudinally continuous models (*5,6*), cellular automata models (*7–9*), and social force models (*10–12*). Longitudinally continuous models determined the longitudinal speeds of vehicles according to the speed and position information of their preceding vehicles. This type of model used a discrete lane selection model to describe the lateral movement of a vehicle. The vehicle took into account the locations and speeds of the surrounding road users and selected a desired route during the lane selection process (*13*). Cellular automata models are a form of discrete time

[1]School of Transportation and Logistics, Southwest Jiaotong University, Chengdu, China
[2] Traffic Management Research Institute of the Ministry of Public Security, Wuxi, China

Corresponding Author:
Address correspondence to Sijing Liu: liusijing666@126.com

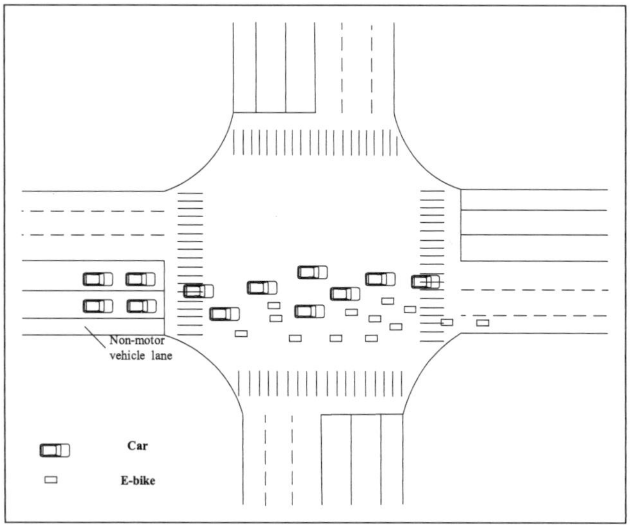

Figure 1. Schematic diagram of heterogeneous traffic flow on signalized intersection.

and space model in which a vehicle passed through the cell grids using some rules, including acceleration rule, deceleration rule, randomization rule, and movement update rule (*14,15*). This type of model needed to separately set different lane-changing rules and update rules for motor vehicles and non-motor vehicles, which increased the complexity of the model. However, the intersection did not have lane division, but both the above two model types assumed lanes exist in the intersection area, which was not consistent with reality. The social force model was first proposed by Helbing and Molnar (*16*) for the movement of pedestrians. The basic principle of the model was that the vehicle was controlled by the sum of several forces acting on the vehicle, such as attracting force, repulsive force, and so forth. As the road user in the social force model could move freely on a two-dimensional plane, the vehicle movement was not constrained by lanes. Thus, the social force models can better describe the behavior of vehicles at an intersection. However, the existing proposed social force models cannot capture the mixture characteristics of the heterogeneous traffic flow of motor vehicles and e-bikes, especially the shuttling behavior of e-bikes, which is as a result of their faster starting velocities. Therefore, a novel social force model is proposed in this paper for the mixed traffic flow of motor vehicles and e-bikes at a signalized intersection.

To calibrate and validate the proposed model, field data from two signalized intersections were collected in Chengdu, China. To reduce the interference of e-bikes with motor vehicles, two strategies are discussed: the stop-line-ahead strategy and the green-signal-ahead strategy. The stop-line-ahead strategy refers to moving the non-motor vehicle stop line forwards for several meters to realize the separation of the two traffic flow in space, and the green-signal-ahead refers to bringing the green signal ahead of non-motor vehicles for several seconds to realize the separation of the two traffic flow in time.

Methodology

Social Force Model for the Heterogeneous Flow of Motor Vehicles and E-Bikes

Driver Behavior Analysis. Because of the invasion of e-bikes, driver behaviors exhibit obviously different characteristics at intersections without invasion of e-bikes. Three types of driver behavior were found: free-driving behavior, following behavior, and shuttling behavior.

Free-driving behavior refers to the driver behavior when motor vehicles or e-bikes can pass the intersection freely without interactions. This type of behavior is found in motor vehicles and e-bikes whose behaviors are not influenced by other vehicles. The speeds of the vehicles are only restricted by the intersection speed limit.

Following behavior means a motor vehicle or an e-bike has to follow its preceding vehicle in passing the intersection, which is the same as car-following behavior in the microscopic traffic flow community.

In shuttling behavior, vehicles are constantly searching for passable gaps to obtain faster speed or a more forward position. The vehicles adopting this type of behavior are faster vehicles. Because of their smaller size and higher acceleration, shuttling behavior is often adopted by e-bikes.

The above analysis indicates that the existing car-following and lane-changing models are not applicable for modeling the behaviors of motor vehicles and e-bikes at an intersection.

Social Forces Acting on Vehicles. According to the above behavior analysis, a novel social force model is proposed for the mixed traffic flow of motor vehicles and e-bikes, in which the behavior is captured by four types of force: self-driven force, following force, passable-gap force, and repulsive force.

Self-Driven Force. This force allows the vehicle to adjust its speed during each adaptation time to reach the maximum speed allowed at the intersection. An acceleration limit is added to the self-driven force to ensure the acceleration of the motor vehicle and the e-bike does not exceed its maximum acceleration. The self-driven force formula is as follows:

$$\vec{F}_j^{\mathrm{dri}} = \min\left\{a_{\max}, v_j^0(t)\vec{e}_j^{\,0} - v_j(t)/\tau_j\right\} \tag{1}$$

where, $a_{\max}$ denotes the maximum acceleration of a vehicle, $v_j^0(t)$ and $v_j(t)$ denote the desired speed and actual speed respectively, τ_j denotes the adaptive time that determines how quickly the vehicle j can change its velocity to the desired velocity, and is a vector, denotes a vehicle's desired direction, and it has the following expression:

$$\vec{e}_j^{\,0}(t) = \frac{x_d(t) - x_0(t)}{\left\|x_d(t) - x_0(t)\right\|} \tag{2}$$

where $x_0(t)$ denotes the current position of the vehicle j, and $x_d(t)$ denotes the desired position of the vehicle j.

Following Force. The following force controls the car-following behavior of vehicles. In traffic, crashes will happen if vehicles touch other vehicles, so a crash-avoiding algorithm is added to the following force to ensure safety. The following force has different signs depending on the relationship between the following vehicle's velocity and a predefined safe velocity. The function of the following force $\vec{F}_i^{\text{fol}}$ is as follows,

$$\vec{F}_i^{\text{fol}} = \min\left\{a_{\max}, v_i^{\text{safe}}(g_i, v_i, v_j) - v_i(t)/\tau_i\right\} \cdot \vec{n}_{\text{i,j}} \quad (3)$$

where $\vec{n}_{\text{i,j}}$ denotes the direction of the following force pointing to its preceding vehicle, and $v_i^{\text{safe}}(t)$ denotes the safe velocity of the vehicle i, which ensures no crashes happen in the following behavior. The safe velocity in Equation (3) adopts Gipps' safe velocity algorithm (*17*).

Passable-Gap Force. The passable-gap force is employed in shuttling behavior. When a passable gap appears, the motor vehicles or e-bikes adopting shuttling behavior will be acted on by the passable-gap force. The force can ensure that the shuttling vehicle can pass the passable gap without crashing with other vehicles. The function of the passable-gap force is as follows:

$$\vec{F}_i^{\text{ge}} = \begin{cases} A^G \exp\left\{\left[d_G - (d_f + 2\Delta d)\right] \Big/ (d_f + 2\Delta d)\right\} \cdot \vec{n}_{\text{ij}} & d_G > d_f + 2\Delta d \\ 0 & d_G \le d_f + 2\Delta d \end{cases} \quad (4)$$

where, d_G denotes the width of the passable gap, d_f denotes the vehicle width of the subject vehicle, Δd denotes the safe side distance from the subject vehicle to the side vehicle when the subject vehicle travels through the passable gap; its value is $\Delta d = d_f/4$, $\vec{n}_{ij} = (r_i - r_j)/d_{ij}$ and denotes the direction of passable-gap force, which is a standard vector pointing from the front midpoint of the subject vehicle i to the midpoint of the passable gap, and A^G is a parameter that needs to be calibrated.

Repulsive Force. For motor vehicles or e-bikes to avoid collisions with lateral vehicles or the boundary, the vehicle must keep a distance from them, and this requirement is met by the repulsive force (*16,18,19*). If the distance from a vehicle to its lateral vehicle or the boundary is within the range of action, the repulsive force decreases with the decrement of distance. Otherwise, the repulsive force is zero. The function of the repulsive force between a vehicle and its lateral vehicle is as follows:

$$\vec{F}_{\text{im}}^{\text{sv}} = \begin{cases} A_1^P \cdot \exp\left[(d_{\text{im}} - B_1)/B_1\right] \cdot \vec{n}_{\text{mi}} & d_{\text{im}} < B_1 \\ 0 & d_{\text{im}} \ge B_1 \end{cases} \quad (5)$$

where, d_{im} denotes the distance from the vehicle i to the vehicle m, B_1 denotes the acting scope of the repulsive force, which means there will be a repulsive force when d_{im} is less than B_1 (here, by observation, B_1 = 0.8 m too), $\vec{n}_{\text{mi}}$ denotes a standard vector pointing from vehicle m to the vehicle i vertically, and A_1^P is a parameter that needs to be calibrated.

Similarly, when the vehicle is too close to a boundary, a repulsive force is also produced. The function of the repulsive force between a vehicle and a boundary is as follows:

$$\vec{F}_{\text{iw}}^{\text{bou}} = \begin{cases} A_2^P \cdot \exp\left[(d_{\text{iw}} - B_2)/B_2\right] \cdot \vec{n}_{\text{wi}} & d_{\text{iw}} < B_2 \\ 0 & d_{\text{iw}} \ge B_2 \end{cases} \quad (6)$$

where d_{iw} denotes the distance from the vehicle i to the boundary w, B_2 denotes the acting scope of the repulsive force, which means there will be a repulsive force when d_{iw} is less than B_2 (here, by observation, B_2 = 0.8 m too), $\vec{n}_{\text{wi}}$ denotes a standard vector pointing from boundary w to the vehicle i vertically, and A_2^P is a parameter that needs to be calibrated.

Resultant Force. If the behavior of a motor vehicle or an e-bike is not influenced by other vehicles, the resultant force acting on the vehicle is as follows:

$$\vec{F} = \vec{F}_j^{\text{dri}} + \vec{F}_{\text{jm}}^{\text{sv}} + \vec{F}_{\text{jw}}^{\text{bou}} \quad (7)$$

If the behavior of a motor vehicle or an e-bike is influenced by other vehicles, both the following force and the passable-gap force may exist at the same time; in this case, the vehicle needs to select the bigger one. The resultant force acting on the vehicle is as follows:

$$\vec{F} = \max(\vec{F}_i^{\text{fol}}, \vec{F}_i^{\text{ge}}) + \vec{F}_{\text{im}}^{\text{sv}} + \vec{F}_{\text{iw}}^{\text{bou}} \quad (8)$$

Boundary Curve Function of the Invasion Behavior of E-Bikes

Observations show that there is a boundary that e-bikes are not willing to cross in the invasion behavior of e-bikes. The boundary curve was found to be a parabola-like curve. However, when the speeds of e-bikes reach the speed limit of the intersection, the speed will not increase further, and the space they occupy will not increase further. At this stage, the boundary curve is a straight line. The whole boundary curve of the invasion behavior of e-bikes is shown in Figure 2.

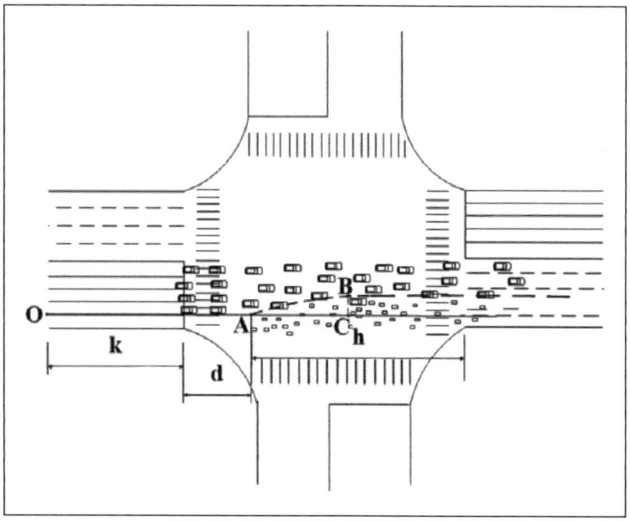

Figure 2. Boundary diagram of e-bikes expanding into driving space of motor vehicles.

In Figure 2, assume the e-bikes start from point A, and reach their maximum velocity at point B. The curve between the two points A and B is a parabolic function in which A is the vertex and B is a point on the parabola. The parabolic function can be uniquely determined by the two points. Take the point O at the stop line for motor vehicles as the origin, that is, O (0, 0), and A (x_a, y_a) and B (x_b, y_b). In Figure 2, the distance from the stop line for motor vehicles to the stop line for e-bikes is denoted as d, so there is A (x_a, y_a) = A (d, 0). Making a plane perpendicular to the OA line through point B can produce point C, so point C is the horizontal position that the leading e-bike can arrive at, at the time of reaching maximum velocity. The distance between A and C can be calculated using the maximum acceleration of e-bikes. The height of BC is related to the number of queuing e-bikes before the stop line, so $BC = v_l^2 / 2a_{\max}$ in which v_l is the speed limit of the intersection. It can be reasonably assumed that the value of BC is in proportion to the number of queuing e-bikes N, that is $y_b = k*N$. Thus, the coordinate of point B is ($v_l^2 / 2a_{\max} + d$, $k*N$).

Assume the function of the parabolic curve is as follows:

$$y = a*\left(x - x_b\right)^2 + b \tag{9}$$

where

$$\begin{cases} a = \left(y_b - b\right) / \left(x_b - x_a\right)^2 \\ b = x_b \end{cases}. \tag{10}$$

Thus, there is,

$$y=\left(k*N-b\right)/\left(v_l^2/2a_{\max}-d\right)^2*\left(x-v_l^2/2a_{\max}\right)^2+v_l^2/2a_{\max} \tag{11}$$

In addition, the straight line starts from point B and is parallel to the lane line on the road, so it also can be determined easily. Therefore, the whole boundary curve of the invasion behavior of e-bikes is obtained.

Experimental Design

The data used in the paper were collected in a metropolis of China, Chengdu city. The data collection time was the evening peak hour (17:30–18:30). Data from two intersections were collected. The basic geometry information of the signalized intersection is presented in Table 1, and the signal cycle and traffic flow information is shown in Table 2.

Model Calibration and Validation

Model Calibration

In the calibration, the data from Site 1 will be used to calibrate the proposed model, and the data from the Site 2 will be used to validate the model. The model calibration process can be abstracted as a nonlinear programming problem. The objective function is the prediction error of the model, as follows:

$$\begin{aligned} &\min\ \omega = \sum_{m=1}^{M} \frac{\left|z_m - \hat{z}_m\right|}{z_m} \\ &s.t. \begin{cases} g_i(para) = 0, i = 1,2,\cdots,M_g \\ h_j(para) \geq 0, j = 1,2,\cdots,M_h \end{cases} \end{aligned} \tag{12}$$

where ω denotes the objective function, M denotes the total number of the observations, z_m denotes the real data for the sample m and has the four alternatives: the total number of motor vehicles passed the intersection in a cycle (N_a), the number of cars passed the intersection in a cycle (N_c), the average passing time of all vehicles (T) and the average speed of all vehicles (V); $\hat{z}_m$ denotes the predicted result for the sample m, $para$ denotes the parameters that need to be calibrated, g_i denotes the ith equality constraints for $Para$, h_j is the jth equality constraints for $Para$, and M_g and M_h denote the number of linear and nonlinear constraints, respectively.

According to the parameter calibration results of the social force model for vehicle dynamics (*19–24*), the feasible ranges of the parameters that need to be calibrated are summarized in Table 3. This study applies the genetic algorithm (GA) (*25*) embedded in Matlab to solve the nonlinear programming problem and derive the optimal model parameters. The optimal parameters obtained are shown in Table 3.

Model Validation

In model validation, the indexes for evaluating the effectiveness of the proposed model are the mean absolute error (MAE), mean absolute relative error (MARE), and Theil's U (*26–28*). The formulas of used indexes are as follows:

Table 1. Geometry Information of the Signalized Intersection

	Site 1		Site 2	
	Direction 1	Direction 2	Direction 1	Direction 2
Number of approach lanes	4	4	4	4
Number of exit lanes	4	4	4	4
Approach lane width (*m*)	3.25	3.25	3.25	3.25
Exit lane width (*m*)	3.50	3.50	3.50	3.50
Non-motor vehicle lane width (*m*)	4.00	4.00	4.00	4.00
Distance between the stop lines of motor vehicles and e-bikes (*m*)	6	6	5	5

Table 2. Signal Cycle and Traffic Flow Information of Signalized Intersection

		Experiment No.	Cars (vehs)	Buses (vehs)	Queuing e-bikes (vehs)	Arrival e-bikes (vehs)	Cycle (s)
Site 1	Direction 1	1	66	0	14	8	60
		2	42	3	17	13	60
		3	39	2	23	22	60
		4	38	1	20	25	60
		Average	46	2	19	17	60
	Direction 2	1	68	2	16	17	80
		2	65	1	21	18	80
		3	63	3	20	17	80
		4	60	3	25	20	80
		Average	64	2	21	18	80
Site 2	Direction 1	1	59	4	18	9	55
		2	53	3	13	11	55
		3	41	2	21	18	55
		4	45	2	24	23	55
		Average	49	3	19	15	55
	Direction 2	1	72	3	20	18	75
		2	70	4	27	24	75
		3	67	2	24	20	75
		4	63	3	26	25	75
		Average	67	3	24	22	75

Table 3. Feasible Ranges of Parameters and Calibration Results of the Social Force Model

Parameters	a_i^{max}	B	τ_i	A^G	A_1^P	A_2^P
Range	0~5	0~3.5	0.5~3.0	0~15	0~15	0~15
Values	3.3	2.9	1.1	3.4	6.3	8.1

$$MAE = \frac{1}{M}\sum_{m=1}^{M} |\hat{z}_m - z_m| \quad (13)$$

$$MARE = \frac{1}{M}\sum_{m=1}^{M}\left(\frac{|\hat{z}_m - z_m|}{z_m}\right) \quad (14)$$

$$U = \frac{\sqrt{\frac{1}{M}\sum_{m=1}^{M}\left(\hat{z}_m - z_m\right)^2}}{\sqrt{\frac{1}{M}\sum_{m=1}^{M}\left(\hat{z}_m\right)^2} + \sqrt{\frac{1}{M}\sum_{m=1}^{M}\left(z_m\right)^2}} \quad (15)$$

After comparing the real data with the simulation results, the model validation results are displayed in Table 4.

Table 4. Model Validation Results

	Direction 1 of Site 1			Direction 2 of Site 1			Direction 2 of Site 2			Direction 2 of Site 2		
Variable	MAE	MARE	U	MAE	MARE	U	MAE	MARE	U	MAE	MARE	U
N_a	3	0.033	0.037	2	0.036	0.037	3	0.036	0.032	2	0.035	0.035
N_c	2	0.010	0.013	2	0.019	0.021	3	0.025	0.014	3	0.022	0.021
T	0.205	0.011	0.013	0.153	0.008	0.008	0.169	0.009	0.014	0.158	0.010	0.009
V	0.5273	0.099	0.089	0.501	0.092	0.095	0.601	0.100	0.085	0.591	0.098	0.093

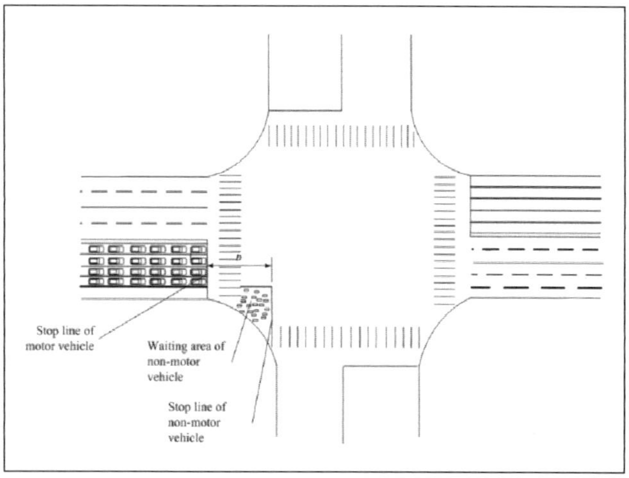

Figure 3. Diagram of the stop-line-ahead strategy.

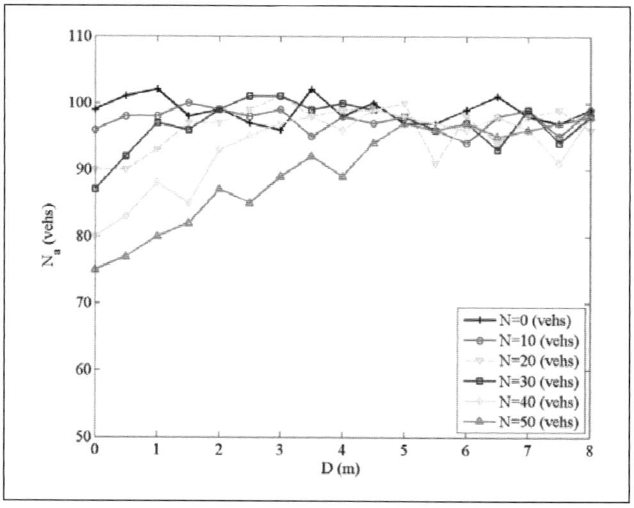

Figure 4. Stop-line-ahead strategy for different numbers of queuing e-bikes.

From Table 4, it can be seen that the prediction errors of the proposed model are <10%, which indicates that the proposed model has high accuracy in replicating the heterogeneous traffic flow of motor vehicles and e-bikes.

Simulations and Results Analysis

Analysis on the Effect of Stop-Line-Ahead Strategy

On the premise of safety, placing the stop line of non-motor vehicles ahead may reduce the invasion effect of e-bikes in motor vehicle traffic, as shown in Figure 3. This strategy can separate the two types of traffic in time to a certain extent. This subsection will investigate the effectiveness of this strategy.

Figure 4 displays the variation of the number of passed vehicles in a cycle with the change of D that is the distance between the two stop lines of motor vehicles and non-motor vehicles. In Figure 4, N is the number of queuing e-bikes waiting after the stop line. From the figure, it can be seen that when the number of queuing e-bikes is small, that is, $N = 0$ or 10, the stop-line-ahead strategy has no obvious effect on the passing efficiency of the intersection. With the increment of number of queuing e-bikes, the effect of stop-line-ahead strategy on improving the efficiency of the intersection becomes more and more significant, especially when $N = 40$ and 50, the number of passed vehicles increases up to 30%. In conclusion, if the space is large enough and safety can be ensured, putting the stop line of non-motor vehicles ahead can effectively reduce the interference of e-bikes in the motor vehicle traffic flow, and further increase the number of passed vehicles in a cycle of the intersection.

Analysis on the Effect of Green-Signal-Ahead Strategy

This subsection evaluates the effect of another possible strategy on reducing the interference of e-bikes, the green-signal-ahead strategy. This strategy activates the green signal for non-motor vehicles several seconds earlier than the green signal for motor vehicles. Unlike the stop-line-ahead strategy, the green-signal-ahead strategy attempts to separate the two types of traffic flow in time. Figure 5 illustrates the evaluation results of the green-signal-ahead strategy. From the figure, similar results to the stop-line-ahead strategy can be seen. When $N = 0$ or 10, that is, the number of the queuing e-bikes is small, there is no need to use the green signal ahead. When $N = 20$ and 30, about 2 s ahead can significantly increase the number of passed vehicles of the intersection, up to 10%. When $N = 40$ and 50, about 4 s ahead will increase to 28% of the number of passed vehicles at the intersection. Therefore,

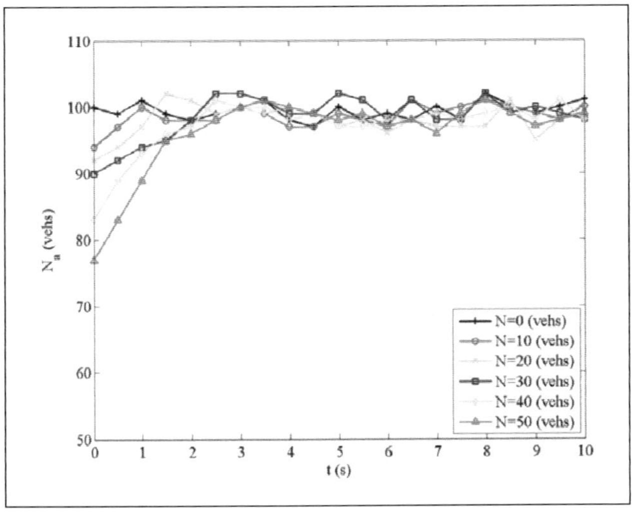

Figure 5. Green-signal-ahead strategy for different numbers of queuing e-bikes.

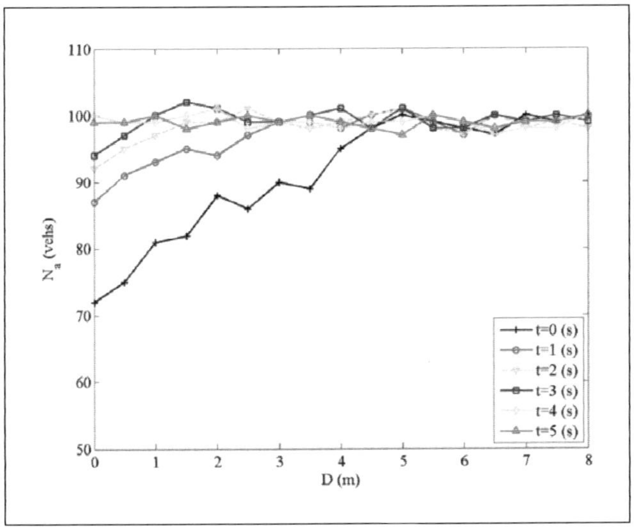

Figure 6. Combination effect of the stop-line-ahead and green-signal-ahead strategies.

bringing the green signal for non-motor vehicles several seconds ahead will significantly improve the efficiency of the intersection, which provides a possible way of reducing the interference of e-bikes on motor vehicle traffic.

Analysis on the Combination Effect of the Stop-Line-Ahead and Green-Signal-Ahead Strategies

This subsection explores the combination effect of the stop-line-ahead strategy and green-signal-ahead strategy on reducing the interference of e-bikes. A scenario where the number of queuing e-bike is large ($N = 50$) is used. As shown in Figure 6, when $t = 2$ s and $D = 2$ m, N_a reaches the maximum value. In Figures 4 and 5, N_a reaches the maximum value at $t = 5$ s or $D = 4$ m, when the stop-line-ahead strategy or the green-signal-ahead strategy is adopted, respectively. This phenomenon indicates that the same control objective can be achieved when adopting the two strategies simultaneously, with a relatively smaller change to the original intersection design.

Conclusions

This paper attempts to investigate the invasion behavior of e-bikes at signalized intersections. A heterogeneous traffic flow model of motor vehicles and e-bikes at urban signal intersections is proposed. The effects of the two strategies on reducing the interference of e-bikes on motor vehicle traffic are evaluated using simulations based on the proposed model. The following conclusions are drawn in this paper.

(1) This paper improves the existing social force model and proposes a new model for the mixed traffic flow of motor vehicles and e-bikes. The model validation results indicate that the model prediction error is small and the proposed model can replicate the traffic flow with high accuracy.
(2) When the number of queuing e-bikes before the stop line is less than 20, the invasion effect of e-bikes on motor vehicles is negligible and there is no need to apply any specific strategy to control the movement of e-bikes.
(3) It is useful to place the stop line of non-motor vehicles ahead, which will separate the mixed traffic flow in space and reduce the interference of e-bikes with motor vehicles. Allowing the stop line ahead of 3–5 meters can increase the volume of motor vehicles in the case where the number of queuing e-bikes is more than 20, up to 30%.
(4) Bringing the green signal of non-motor vehicle ahead can also reduce the interference of e-bikes with motor vehicles. When the number of queuing e-bikes is more than 20, 2–4 s ahead of green signal can increase up to 30% percent of traffic volume of motor vehicles.
(5) The same control objective can be achieved when combining the two strategies together, and it need a relatively smaller change to the original intersection design.

Acknowledgments

This work was supported by National Science Foundation of China (Grant No. 51408509, Grant No. 71603219) and Fundamental Research Funds for the Central Universities (Grant No. 2682016CX046, Grant No. 2682016CX051).

References

1. Weinert, J., C. Ma, X. Yang, and C. Cherry. Electric Two-Wheelers in China: Effect on Travel Behavior, Mode Shift,

and User Safety Perceptions in a Medium-Sized City. *Transportation Research Record: Journal of the Transportation Research Board*, 2007. 2038: 62–68.
2. Chen, X., C. Shao, and H. Yue. Influence of Bicycle Traffic on Capacity of Typical Signalized Intersection. *Tsinghua Science & Technology*, Vol. 12, No. 2, 2007, pp. 198–203.
3. Phillips, R. O., T. Bjørnskau, R. Hagman, and F. Sagberg. Reduction in Car–Bicycle Conflict at a Road–Cycle Path Intersection: Evidence of Road User Adaptation? *Transportation Research Part F: Traffic Psychology and Behaviour*, Vol. 14, No. 2, 2011, pp. 87–95.
4. Allen, D., J. Hummer, N. Rouphail, and J. Milazzo. Effect of Bicycles on Capacity of Signalized Intersections. *Transportation Research Record: Journal of the Transportation Research Board*, 1998. 1646: 87–95.
5. Oketch, T. New Modeling Approach for Mixed-Traffic Streams with Nonmotorized Vehicles. *Transportation Research Record: Journal of the Transportation Research Board*, 2000. 1705: 61–69.
6. Mathew, T. V., O. Abdullah, C. R. Munigety, and S. Anirudha. Strip-Based Simulation Model for Mixed Traffic Conditions. Presented at 91st Annual Meeting of the Transportation Research Board, Washington, D.C., 2012.
7. Vasic, J., and H. J. Ruskin. Cellular Automata Simulation of Traffic Including Cars and Bicycles. *Physica A: Statistical Mechanics and its Applications*, Vol. 391, No. 8, 2012, pp. 2720–2729.
8. Shunping, J., P. Hongqin, G. Jinyi, and C. Haibo. Quantitative Analysis of Impact of Bicycles on Vehicles in Urban Mixed Traffic. *Journal of Transportation Systems Engineering and Information Technology*, Vol. 8, No. 2, 2008, pp. 58–63.
9. Jiang, R., B. Jia, and Q. S. Wu. Stochastic Multi-Value Cellular Automata Models for Bicycle Flow. *Journal of Physics A: Mathematical and General*, Vol. 37, No. 6, 2004, p. 2063.
10. Li, M., F. Shi, and D. Chen. Analyze Bicycle-Car Mixed Flow by Social Force Model for Collision Risk Evaluation. Presented at 3rd International Conference on Road Safety and Simulation, Indianapolis, IN, 2011.
11. Liang, X., M. Baohua, and X. Qi. Psychological-Physical Force Model for Bicycle Dynamics. *Journal of Transportation Systems Engineering and Information Technology*, Vol. 12, No. 2, 2012, pp. 91–97.
12. Schönauer, R., M. Stubenschrott, W. Huang, C. Rudloff, and M. Fellendorf. Modeling Concepts for Mixed Traffic: Steps Toward a Microscopic Simulation Tool for Shared Space Zones. *Transportation Research Record: Journal of the Transportation Research Board*, 2012. 2316: 114–121.
13. Twaddle, H., T. Schendzielorz, and O. Fakler. Bicycles in Urban Areas: Review of Existing Methods for Modeling Behavior. *Transportation Research Record: Journal of the Transportation Research Board*, 2014. 2434: 140–146.
14. Yao, D., Y. Zhang, L. Li, Y. Su, S. Cheng, and W. Xu. Behavior Modeling and Simulation for Conflicts in Vehicles-Bicycles Mixed Flow. *IEEE Intelligent Transportation Systems Magazine*, Vol. 1, No. 2, 2009, pp. 25–30.
15. Mallikarjuna, C., and K. R. Rao. Cellular Automata Model for Heterogeneous Traffic. *Journal of Advanced Transportation*, Vol. 43, No. 3, 2009, pp. 321–345.
16. Helbing, D., and P. Molnar. Social Force Model for Pedestrian Dynamics. *Physical Review E(S1539-3755)*, Vol. 51, No. 5, 1995, pp. 4282–4285.
17. Gipps, P. G. A Behavioural Car-Following Model for Computer Simulation. *Transportation Research Part B: Methodological*, Vol. 15, No. 2, 1981, pp. 105–111.
18. Huang, W., M. Fellendorf, and R. Schönauer. Social Force Based Vehicle Model for 2-Dimensional Spaces. Presented at 91st Annual Meeting of the Transportation Research Board, Washington, D.C., 2011.
19. Huynh, D. N., M. Boltze, and A. T. Vu. Modelling Mixed Traffic Flow at Signalized IntersectionUsing Social Force Model. *Journal of the Eastern Asia Society for Transportation Studies*, Vol. 10, 2013, pp. 1734–1749.
20. Anvari, B., M. G. Bell, A. Sivakumar, and W. Y. Ochieng. Modelling Shared Space Users Via Rule-Based Social Force Model. *Transportation Research Part C: Emerging Technologies*, Vol. 51, 2015, pp. 83–103.
21. Chen, D. *Modeling and Analyzing of Mixed Traffic Flow on the Urban Road Section Based on Social Force*. Thesis. Central South University, Wu Han., 2011.
22. Fei, L., H. Zhu, and X. Han. Analysis of Traffic Congestion Induced by the Work Zone. *Physica A: Statistical Mechanics and its Applications*, Vol. 450, 2016, pp. 497–505.
23. Helbing, D., and B. Tilch. Generalized Force Model of Traffic Dynamics. *Physical Review E*, Vol. 58, No. 1, 1998, pp. 133–138.
24Gunawan, F. E. Two-Vehicle Dynamics of the Car-Following Models on Realistic Driving Condition. *Journal of Transportation Systems Engineering and Information Technology*, Vol. 12, No. 2, 2012, pp. 67–75.
25. Houck, C. R., J. Joines, and M. G. Kay. A Genetic Algorithm for Function Optimization: A Matlab Implementation. *Ncsu-ie tr*, Vol. 95, No. 09, 1995, pp. 1–10.
26. Meng, Q., and J. Weng. An Improved Cellular Automata Model for Heterogeneous Work Zone Traffic. *Transportation Research Part C: Emerging Technologies*, Vol. 19, No. 6, 2011, pp. 1263–1275.
27Punzo, V., and F. Simonelli. Analysis and Comparison of Microscopic Traffic Flow Models with Real Traffic Microscopic Data. *Transportation Research Record: Journal of the Transportation Research Board*, 2005. 1934: 53–63.
28. Tan, H., Y. Wu, B. Shen, P. J. Jin, and B. Ran. Short-Term Traffic Prediction Based on Dynamic Tensor Completion. *IEEE Transactions on Intelligent Transportation Systems*, Vol. 17, No. 8, 2016, pp. 2123–2133.

The Standing Committee on Transportation in the Developing Countries (ABE90) peer-reviewed this paper (18-02478).

Article

TRR
JOURNAL OF THE TRANSPORTATION RESEARCH BOARD

Transportation Research Record
2018, Vol. 2672(31) 41–53

DOI: 10.1177/0361198118758684
journals.sagepub.com/home/trr
SAGE

Influence of Socioeconomic Conditions on Crash Injury Severity for an Urban Area in a Developing Country

Rodrigo Mesa-Arango[1], Víctor G. Valencia-Alaix[2], Raul A. Pineda-Mendez[3], and Taleb Eissa[1]

Abstract

This paper includes macroeconomic conditions in an econometric framework to understand urban crash injury severity (CIS) in a developing country, and identify its distinctive socioeconomic conditions. The work combines classic variables from a unique data set of crashes in Medellín, Colombia, with macroeconomic indicators. A multinomial logit (MNL) model with random parameters mines valuable information from the data. Numerical results support the following CIS mitigation policies: upgrading intersections with traffic signals; incorporating forgiving roadway designs; providing better conditions for motorcyclists and non-motorized users; prioritizing education, outreach, and enforcement campaigns during periods of good macroeconomic conditions (for some segments of the population), high motorization rates, and regarding specific periods, that is, times within the day, the week, and the year.

This paper is one of the few studies on crash injury severity (CIS) in urban areas in developing countries. To our knowledge, it is the first to employ advanced econometric models for CIS in this context, and to link national macroeconomic conditions and CIS for disaggregated safety analysis in urban areas. The study was conducted in the city of Medellín, the second largest city in the developing country of Colombia.

Traffic safety is a global concern. According to the World Health Organization (WHO), more than 1.25 million fatalities per year can be attributed to traffic safety issues in developing/low-income countries (*1*). In 2016, 6,925 road traffic deaths occurred in Colombia (*2*), costing approximately 3.6% of Colombia's gross domestic product (GDP) (*3*). Colombia receives a poor grade of 2 out of 10 based on the WHO scale of road safety. Critical issues include deficiencies in implementing speed limits and seatbelt usage. The most vulnerable users in Colombia are motorcyclists (44% of crashes) and pedestrians (29% of crashes).

In developed countries numerous road safety studies are conducted, with rigorous methodologies using rich data sets, to enhance understanding of crashes, to foresee issues, and to mitigate them with improved roadway design, education, and outreach. Conversely, developing countries face challenges that reduce the resources for such studies. Mitigation campaigns tend to rely on qualitative approaches, for example, expert opinions, which might be cheap but not cost effective because biased results can aggravate safety issues. Since the world is moving rapidly toward big data availability and business analytics, new data sets are becoming available in developing nations and advanced methodologies can significantly benefit safety studies and broaden our global understanding of traffic safety.

Road safety is traditionally studied from two perspectives: crash frequency, that is, number of crashes per unit of time and geography, and CIS, that is, propensity to suffer different levels of severity resulting from crashes. Although both are equally important, this research focuses on CIS, which is approached from two econometric perspectives. First, ordered outcome models (e.g., ordered probit and variations) assume ordinal CIS outcomes, which is an intuitive supposition but requires covariates to monotonically affect the dependent variables. Unfortunately, this behavior does not necessarily occur in all CIS situations, that is, the probability of suffering intermediate categories can be more likely for some variables. The second approach, discrete outcome models, overcomes this issue and is employed in this study. A multinomial logit (MNL) model with random parameters is estimated to

[1]Department of Civil Engineering and Construction Management, Florida Institute of Technology, Melbourne, FL
[2]Ingeniería Civil - Facultad de Minas, Universidad Nacional de Colombia (Medellín), Medellín, Antioquia, Colombia
[3]Lyles School of Civil Engineering, Purdue University, West Lafayette, IN

Corresponding Author:
Address correspondence to Rodrigo Mesa Arango: rmesaarango@fit.edu

analyze urban CIS in a developing country. There are few econometric models for CIS in developing countries. Therefore, the first contribution of this paper is to develop the first MNL with random parameters to understand urban CIS in a developing nation, using data from Medellín, Colombia.

The paper identifies the following covariate groups used to explain CIS in developed countries: (*a*) characteristics of the users, (*b*) behavior of the users, (*c*) environmental conditions, (*d*) temporal attributes, (*e*) transportation infrastructures, and (*f*) crash-specific attributes. Recognizing them is important to create a developing-country model that is consistent with previous works and the available data. Likewise, it helps test the explanatory power of the proposed macroeconomic covariates rigorously.

Cultural, sociodemographic, and idiosyncratic conditions distinctively differentiate developed and developing countries. The key challenge in constructing the CIS model for a developing country is to incorporate such behavioral components correctly. This research hypothesizes macroeconomic variables related to GDP per capita, consumer price index (CPI), and motorization rates to approximate the connection between unique unobserved behaviors and CIS. The intuition, significance, and impact of all the variables is tested via model development. Since no paper has tested the macroeconomic conditions that affect CIS in developing countries using an econometric framework, the second contribution of this paper is to demonstrate that macroeconomic variables indeed affect urban CIS using a MNL model with random parameters.

Development of the model was possible thanks to a unique data set compiled by the Medellín Mobility Agency from police reports between 2009 and 2016. Medellín is one of the top three cities in Colombia with respect to area, population, and economic activities. The city hosts 2.5 million people in a 382-square-kilometer area. Medellín is the core city of the Metropolitan Area of the Aburra Valley (composed of 10 municipalities). Its location near the Equator line generates constant sunlight over the year between 6:00 a.m and 6:00 p.m. (no seasons). The average altitude is 1,538 meters above sea level, the average temperature is 72.5 degrees Fahrenheit, and it rains on average 227 days during the year. Medellín is located in a U-shaped valley crossed by the Medellín river (flowing north), with level terrains around the river and rolling/mountainous conditions toward the hills (east and west). The crash data set is unique for Colombia, where transportation data is generally unavailable for researchers. With 357,696 observations (crashes), broad geographic coverage, long timespan (8 years), and detailed information, the data set is extremely advantageous for CIS analysis.

The numerical results indicate the effect of several classic covariates on CIS in this region, for example, variables related to the roadway and traffic control, temporal attributes, crash-specific attributes, and environmental conditions. Macroeconomic indicators, that is, GDP per capita, CPI, and motorization, also explain CIS, which links the unique socioeconomic context of the developing nation with urban CIS. Thus, the third contribution of the paper is to provide key findings to influence policy-making and improve CIS. The following actions are proposed: enhance intersections with traffic signals, implement forgiving roadway designs, improve conditions for motorcyclists and nonmotorized users, and implement safety campaigns during certain periods (related to macroeconomic conditions and time of day/week/year).

The paper is organized as follows. The introduction contextualizes and justifies the investigation. The literature review examines methodological approaches, modeling trends, and explanatory variables that illustrate gaps in the literature. The Data section explains the corresponding data set and summary statistics. Econometric Approach describes the methodology of the MNL with random parameters. Estimation Results summarizes and discusses the numerical results as well as policy implications. The conclusion recapitulates the work and provides future research directions.

Literature Review

Due to the vast amount of works on CIS available in the literature, this review prioritizes recent research (after 2010), and/or frequently cited papers in high-impact-factor journals (around 1.5 and 3.0). The following subsections expand on econometric approaches, CIS variables, and gaps in the literature.

Key Econometric Approaches to Modeling CIS

CIS is studied via discrete outcome models. The two main econometric approaches are ordered and unordered outcome paradigms. Ordered outcome models are popular because CIS is an ordinal variable where fatal crashes are more severe than crashes with injuries, which are more severe than property damage only (PDO) crashes. Multiple works follow this approach (e.g., [*4–12*]). However, in many cases unordered discrete-choice models are more suitable because CIS does not strictly increase/decrease monotonically with respect to all explanatory variables, for example, airbag deployment increases the probability of injuries (intermediate outcome) and reduces the probability of PDO and fatalities (see Section 14.3 in [*13*] for more details). Thus, many studies employ unordered models to understand CIS (e.g., [*5, 7, 14–30*]). Since unordered outcome models provide a flexible framework to understand CIS, they are selected for the current study.

Main CIS Covariates

Table 1 summarizes and categorizes key explanatory variables used to model CIS in the past. The first column describes variables. The second column classifies variables into six groups, that is, A to F (explained below). The third column counts the number of papers related to each variable. Finally, the fourth column provides the corresponding citation.

Table 1. Variables Used in Previous CIS Research

Variable	Group	Times used	References
Gender	A	24	(4–11, 14–29)
Age	A	22	(4–11, 14–24, 26, 28, 29)
Vehicle type	A	22	(4–12, 14, 15, 17–19, 21–28)
Model	A	10	(5, 6, 8, 9, 15, 18, 21, 23, 24, 26)
Alcohol	B	17	(4, 6–8, 11, 14, 16–26)
Seatbelt	B	12	(4, 5, 7–9, 17–19, 21, 24–26)
Speeding	B	12	(6–8, 14–18, 20, 23, 26, 29)
Drug	B	11	(8, 14, 16, 17, 19–25)
Weather	C	20	(4, 5, 7, 9–11, 14–19, 21–27, 29)
Light	C	19	(4–7, 9, 11, 14–23, 25, 26, 29)
Surface	C	13	(5, 7, 9, 11, 14–27, 29)
Time of day	D	18	(5, 7–10, 12, 15, 17–20, 22, 24–29)
Day of week	D	13	(5–8, 15–18, 20, 22, 23, 25, 28)
Month	D	8	(5, 7, 15, 17, 20, 22, 23, 29)
Holiday	D	2	(7, 17)
Speed limit	E	17	(4–9, 14–17, 20–24, 26, 28)
Functional class	E	14	(5, 7, 9, 14–22, 24, 26)
Grade	E	13	(4, 7, 8, 11, 14, 16–18, 21, 22, 25, 26, 29)
Lanes	E	11	(4, 6, 8, 14, 15, 18, 21, 24–27)
Curve	E	10	(4, 6, 7, 11, 16–18, 21, 25, 26, 29)
Traffic control	E	10	(5, 7–9, 16–19, 24, 25)
Pavement	E	8	(7, 8, 10, 14, 16, 17, 19, 25)
Land use	E	7	(8, 12, 15, 18, 23, 26, 28)
Condition	E	5	(16, 18, 19, 21, 26)
Direction	E	2	(6, 26)
Crosswalk	E	1	(16)
Collision type	F	14	(4, 6, 8, 9, 11, 15, 17–19, 21, 23, 24, 26, 29)
Obstacle	F	11	(4, 7, 9, 15, 17–21, 24, 26)
Number of vehicles	F	5	(4, 6, 8, 15, 25)
Overturn/rollover	F	5	(4, 11, 17, 20, 24)

Note: A = characteristics of the users involved in the crash; B = behavior of the users involved in the crash; C = environmental conditions of the crash; D = temporal attributes of the crash; E = transportation infrastructure associated with the crash; F = crash-specific attributes.

The key variables explaining CIS can be classified into the following six groups: (*a*) User characteristics, that is, age, gender, and vehicle type; (*b*) user behavior, that is, alcohol/drug consumption, seatbelt usage, and speeding; (*c*) environmental conditions, that is, light, weather, and surface conditions; (*d*) temporal attributes, that is, time of day and day of the week; (*e*) transportation infrastructure, that is, speed limit, functional class, curves, among others; and (f) crash-specific attributes, that is, collision type, obstacle, and number of vehicles.

Trends and Gaps in the Literature

Previous works focus their attention on very specific types of crashes. Thus, several authors focus on single-vehicle crashes (*9, 17, 19–21, 26*), two-vehicle crashes (*5, 7, 23*), pedestrian crashes (*22, 27, 28*), specific types of vehicles like motorcycles (*20, 23*) or heavy-vehicles (*12, 19, 29*), and contributing/aggravating attributes from transportation infrastructure and driver behavior (*18*). However, few papers combine multiple types of crashes to achieve holistic conclusions and find the common explanatory power of variables across crash types. The current research will follow that approach which facilitates application in system-wide decision-making processes.

Additionally, previous works targeted very specific geographic contexts to understand CIS. Several authors investigate basic roadway segments, that is, without intersections (*17, 22*). Other papers study just crashes at intersections (*6, 22*). Other researchers analyze either rural or urban contexts (*5, 9, 19, 21, 27–29*). Thus, this paper will be one of the few combining multiple geographic contexts to understand CIS in an urban context with a core city surrounded by small municipalities which is common in developing countries.

Furthermore, CIS analysis in developing countries is a relatively new area. A few recent studies (*27, 28, 31, 32*) address CIS in developing countries, mostly to understand vulnerable users, such as pedestrians and cyclists. Additionally, one of the first papers aiming to introduce macroeconomic conditions in CIS studies is Bishai (*33*). That investigation constructs a country-level fixed-parameter model for 41 nations (developed and developing) and finds GDP per capita to increase CIS in urban areas. Therefore, the current paper narrows these gaps by developing a holistic CIS study that jointly captures multiple crash types, geographic contexts, and macroeconomic conditions in the context of a metropolitan area in a developing country, the city of Medellín in Colombia. More details on the data and context of the study are provided next.

Data

This study employs 357,696 observations of crashes compiled by the Medellín Mobility Agency from police reports between 2009 and 2016. Injuries and fatalities involved 616,538 vehicles and about 226,211 individuals. Table 2 provides summary statistics for selected variables of the data set.

Three severity levels are recorded, that is, PDO (49.16%), injury (50.52%), and fatality (0.33%). Since fatal observations are marginal, they are aggregated with injuries into an injury-and-fatal outcome (IFO) category.

The following trends are observed in the data set:

- Roadway design and traffic control: Most crashes occur on curves (92%), single carriageways (76%), and unidirectional roads (69%); some on two-lane roads (48%), one-lane roads (34%), and intersections (14%); few occur on roundabouts (2%).
- Pavement conditions: Most crashes occur on dry pavements (93%).
- Temporal attributes: Yearly/monthly crashes remain relatively constant. More crashes occur on Fridays (15%) and fewer on Sundays (10%). The daily crash distribution is midnight to 6:00 a.m.: 9%, 6:00 a.m to midday: 31%, midday to 6:00 p.m: 40%, and 6:00 p.m to midnight: 21%.
- Crash specific attributes: 68% of the records are regular crashes, 11% run overs, 10% other uncommon circumstances, in 9% occupants fall from vehicles and 2% are rollovers.
- Environmental attributes: About 6% of the crashes are related to rain, fog, and/or wind. The land uses associated with the crashes are residential (59%), commercial (35%), and industrial (5%).

The following macroeconomic indicators are compiled from the World Bank's databank (*34*) and summarized in Table 2.

- GDP per capita: Yearly GDP divided by population. Evaluates national economic growth and population productivity (average 6,788 U.S. dollars).
- CPI: A yearly weighted average of prices for a basket of goods and services. Evaluates the average cost of living (9.1% average increase with 2010 as a base).
- Motorization: New vehicle sales per 1,000 people. Evaluates distribution of new vehicles with respect to population (6.0 on average).

Some limitations of the data set include the presence of missing and inconsistent values, that is, about 14.3% of the data in the final model estimation. The owner pre-processed the data for quality assurance and quality control (QA/QC), which addressed most of its limitations.

The method used to analyze the data set is presented next.

Econometric Approach

CIS is analyzed with a discrete choice modeling framework, that is, a MNL with random parameters (*35*), which can consider random severity propensity variation, and overcomes problems related to correlation of unobserved factors, and restrictive substitution patterns of the regular MNL.

For CIS, let U_{ni} be the disutility of a crash $n \in N$ suffering the most severe type $i \in I$, where I is the set of injury types, and N the data set of crashes. Equation 1 formulates U_{ni}, where X_{ni} is a vector of crash- and outcome-specific variables, β a vector of estimable parameters, and ε_{ni} a random term (iid extreme value).

$$U_{ni} = \beta' X_{ni} + \varepsilon_{ni}, \forall i \in I, \forall n \in N \tag{1}$$

Equation 2 describes the probability $L_{ni}(\beta)$ of crash $n \in N$ suffering severity type $i \in I$ conditional on β.

$$L_{ni}(\beta) = \frac{e^{\beta' X_{ni}}}{\sum_{j \in I} e^{\beta' X_{nj}}}, \forall i \in I, \forall n \in N \tag{2}$$

For random parameters, the unconditional probability P_{ni} (Equation (3) integrates the product over all values of β according to its continuous density function $f(\beta)$: normal, lognormal, uniform, among others.

$$P_{ni} = \int L_{ni}(\beta) f(\beta) d\beta \tag{3}$$

Model estimation involves computing $L_{ni}(\beta)$ with draws of β generated from $f(\beta)$ for a sufficient number of draws. Results are averaged to simulate P_{ni} and compute the corresponding likelihood function LL (Equation 4). The best estimation of β maximizes LL. $\delta_{in} = 1$ in Equation 4 if the observed outcome of $n \in N$ is $i \in I$, $\delta_{in} = 0$ otherwise. Bhat (*36*) and Train (*37*) indicate Halton draws to be more efficient than purely random draws for model estimation.

$$LL = \sum_{n \in N} \sum_{i \in I} \delta_{in} \ln(P_{ni}) \tag{4}$$

Table 2. Summary Statistics for Selected Variables in the Data Set

	Total	2009		2010		2011		2012		2013		2014		2015		2016	
Variable	Mean	Mean	Std. D.	Mean	Std. D.	Mean	Std. D.	Mean	Std. D.	Mean	Std. D.	Mean	Std. D.	Mean	Std. D.	Mean	Std. D.
Crash injury severity																	
PDO	4.9E−01	6.2E−02	2.4E−01	6.1E−02	2.4E−01	6.3E−02	2.4E−01	6.7E−02	2.5E−01	6.6E−02	2.5E−01	5.8E−02	2.3E−01	5.8E−02	2.3E−01	5.7E−02	2.3E−01
Injury	5.1E−01	4.8E−02	2.1E−01	4.8E−02	2.1E−01	5.3E−02	2.2E−01	6.4E−02	2.4E−01	7.3E−02	2.6E−01	7.2E−02	2.6E−01	7.3E−02	2.6E−01	7.5E−02	2.6E−01
Fatal	3.3E−03	4.2E−04	2.0E−02	3.7E−04	1.9E−02	4.6E−04	2.2E−02	4.3E−04	2.1E−02	4.2E−04	2.0E−02	3.7E−04	1.9E−02	4.1E−04	2.0E−02	3.9E−04	2.0E−02
Roadway geometry and traffic control																	
One lane	3.4E−01	2.9E−02	1.7E−01	1.7E−02	1.3E−01	2.8E−02	1.7E−01	4.3E−02	2.0E−01	4.9E−02	2.2E−01	6.4E−02	2.4E−01	4.6E−02	2.1E−01	6.4E−02	2.4E−01
Two lanes	4.8E−01	5.4E−02	2.3E−01	6.1E−02	2.4E−01	6.0E−02	2.4E−01	6.2E−02	2.4E−01	6.7E−02	2.5E−01	5.5E−02	2.3E−01	6.7E−02	2.5E−01	5.6E−02	2.3E−01
One carriageway	7.6E−01	8.3E−02	2.8E−01	7.7E−02	2.7E−01	8.6E−02	2.8E−01	1.0E−01	3.0E−01	1.1E−01	3.1E−01	1.0E−01	3.1E−01	9.7E−02	3.0E−01	1.0E−01	3.0E−01
Curve	9.2E−01	9.9E−02	3.0E−01	9.7E−02	3.0E−01	1.0E−01	3.1E−01	1.2E−01	3.2E−01	1.3E−01	3.4E−01	1.2E−01	3.3E−01	1.2E−01	3.3E−01	1.2E−01	3.3E−01
Roundabout	2.2E−02	2.7E−03	5.2E−02	2.8E−03	5.2E−02	3.0E−03	5.5E−02	3.2E−03	5.7E−02	3.1E−03	5.5E−02	2.2E−03	4.7E−02	2.3E−03	4.8E−02	2.8E−03	5.3E−02
Intersection	1.4E−01	1.9E−02	1.4E−01	1.8E−02	1.3E−01	1.9E−02	1.4E−01	2.0E−02	1.4E−01	1.7E−02	1.3E−01	1.6E−02	1.2E−01	1.7E−02	1.3E−01	1.8E−02	1.3E−01
Unidirectional	6.9E−01	7.6E−02	2.7E−01	7.1E−02	2.6E−01	7.9E−02	2.7E−01	9.2E−02	2.9E−01	9.9E−02	3.0E−01	1.0E−01	3.0E−01	8.3E−02	2.8E−01	8.5E−02	2.8E−01
Pavement conditions																	
Dry pavement	9.3E−01	9.8E−02	3.0E−01	9.3E−02	2.9E−01	1.0E−01	3.0E−01	1.2E−01	3.3E−01	1.3E−01	3.4E−01	1.3E−01	3.3E−01	1.3E−01	3.4E−01	1.3E−01	3.4E−01
Temporal attributes																	
January	7.2E−02	8.1E−03	9.0E−02	7.6E−03	8.7E−02	8.1E−03	8.9E−02	8.9E−03	9.4E−02	1.1E−02	1.0E−01	9.7E−03	9.8E−02	9.3E−03	9.6E−02	9.3E−03	9.6E−02
February	7.7E−02	8.5E−03	9.2E−02	8.4E−03	9.1E−02	8.7E−03	9.3E−02	9.6E−03	9.7E−02	1.0E−02	1.0E−01	1.0E−02	1.0E−01	1.0E−02	1.0E−01	1.1E−02	1.0E−01
March	8.6E−02	9.7E−03	9.8E−02	9.8E−03	9.9E−02	9.9E−03	9.9E−02	1.1E−02	1.0E−01	1.1E−02	1.0E−01	1.2E−02	1.1E−01	1.1E−02	1.1E−01	1.1E−02	1.0E−01
April	8.1E−02	9.0E−03	9.5E−02	8.5E−03	9.2E−02	8.8E−03	9.4E−02	9.8E−03	9.9E−02	1.2E−02	1.1E−01	1.1E−02	1.0E−01	1.1E−02	1.0E−01	1.1E−02	1.1E−01
May	8.6E−02	9.6E−03	9.8E−02	8.9E−03	9.4E−02	1.0E−02	9.9E−02	1.1E−02	1.1E−01	1.2E−02	1.1E−01	1.1E−02	1.1E−01	1.2E−02	1.1E−01	1.1E−02	1.1E−01
June	8.0E−02	8.5E−03	9.2E−02	8.5E−03	9.2E−02	9.3E−03	9.6E−02	1.0E−02	1.0E−01	1.1E−02	1.1E−01	1.1E−02	1.0E−01	1.0E−02	1.0E−01	1.1E−02	1.0E−01
July	8.5E−02	9.4E−03	9.7E−02	8.9E−03	9.4E−02	9.7E−03	9.8E−02	1.1E−02	1.0E−01	1.2E−02	1.1E−01	1.1E−02	1.1E−01	1.2E−02	1.1E−01	1.1E−02	1.0E−01
August	8.8E−02	9.4E−03	9.6E−02	9.3E−03	9.6E−02	1.0E−02	1.0E−01	1.2E−02	1.1E−01	1.2E−02	1.1E−01	1.1E−02	1.1E−01	1.2E−02	1.1E−01	1.2E−02	1.1E−01
September	8.7E−02	9.3E−03	9.6E−02	9.6E−03	9.7E−02	1.0E−02	1.0E−01	1.2E−02	1.1E−01	1.2E−02	1.1E−01	1.1E−02	1.0E−01	1.2E−02	1.1E−01	1.1E−02	1.1E−01
October	8.8E−02	9.7E−03	9.8E−02	1.0E−02	1.0E−01	1.0E−02	1.0E−01	1.2E−02	1.1E−01	1.2E−02	1.1E−01	1.1E−02	1.0E−01	1.1E−02	1.0E−01	1.2E−02	1.1E−01
November	8.4E−02	9.0E−03	9.4E−02	9.7E−03	9.8E−02	1.0E−02	1.0E−01	1.1E−02	1.1E−01	1.2E−02	1.1E−01	1.0E−02	1.0E−01	1.0E−02	1.0E−01	1.1E−02	1.0E−01
December	8.7E−02	9.7E−03	9.8E−02	9.9E−03	9.9E−02	1.1E−02	1.0E−01	1.2E−02	1.1E−01	1.2E−02	1.1E−01	1.1E−02	1.0E−01	1.1E−02	1.0E−01	1.1E−02	1.0E−01
Monday	1.4E−01	1.5E−02	1.2E−01	1.5E−02	1.2E−01	1.6E−02	1.3E−01	1.8E−02	1.3E−01	2.0E−02	1.4E−01	1.9E−02	1.4E−01	1.9E−02	1.4E−01	1.9E−02	1.4E−01
Tuesday	1.5E−01	1.6E−02	1.3E−01	1.6E−02	1.3E−01	1.8E−02	1.3E−01	2.0E−02	1.4E−01	2.2E−02	1.5E−01	2.0E−02	1.4E−01	2.1E−02	1.4E−01	2.0E−02	1.4E−01
Wednesday	1.5E−01	1.6E−02	1.2E−01	1.6E−02	1.3E−01	1.7E−02	1.3E−01	1.9E−02	1.4E−01	2.0E−02	1.4E−01	2.0E−02	1.4E−01	2.0E−02	1.4E−01	2.0E−02	1.4E−01
Thursday	1.5E−01	1.6E−02	1.3E−01	1.6E−02	1.2E−01	1.7E−02	1.3E−01	1.9E−02	1.4E−01	2.0E−02	1.4E−01	1.9E−02	1.4E−01	2.0E−02	1.4E−01	2.0E−02	1.4E−01
Friday	1.6E−01	1.7E−02	1.3E−01	1.8E−02	1.3E−01	1.9E−02	1.4E−01	2.1E−02	1.4E−01	2.2E−02	1.5E−01	2.1E−02	1.4E−01	2.0E−02	1.4E−01	2.1E−02	1.4E−01
Saturday	1.5E−01	1.7E−02	1.3E−01	1.6E−02	1.3E−01	1.8E−02	1.3E−01	2.0E−02	1.4E−01	2.0E−02	1.4E−01	1.9E−02	1.4E−01	1.8E−02	1.3E−01	1.9E−02	1.4E−01
Sunday	1.0E−01	1.2E−02	1.1E−01	1.2E−02	1.1E−01	1.2E−02	1.1E−01	1.4E−02	1.2E−01	1.4E−02	1.2E−01	1.3E−02	1.1E−01	1.3E−02	1.1E−01	1.3E−02	1.1E−01

(continued)

Table 2. (continued)

	Total	2009		2010		2011		2012		2013		2014		2015		2016	
Variable	Mean	Mean	Std. D.	Mean	Std. D.	Mean	Std. D.	Mean	Std. D.	Mean	Std. D.	Mean	Std. D.	Mean	Std. D.	Mean	Std. D.
Crash specific attributes																	
Run over	1.1E−01	1.3E−02	1.1E−01	1.2E−02	1.1E−01	1.3E−02	1.1E−01	1.3E−02	1.1E−01	1.4E−02	1.2E−01	1.4E−02	1.2E−01	1.3E−02	1.2E−01	1.2E−02	1.1E−01
Fall from car	8.9E−02	1.1E−02	1.0E−01	7.5E−03	8.6E−02	8.4E−03	9.1E−02	1.2E−02	1.1E−01	1.3E−02	1.2E−01	1.3E−02	1.1E−01	1.2E−02	1.1E−01	1.2E−02	1.1E−01
Crash	6.8E−01	8.1E−02	2.7E−01	8.0E−02	2.7E−01	8.3E−02	2.8E−01	9.0E−02	2.9E−01	9.1E−02	2.9E−01	8.5E−02	2.8E−01	8.7E−02	2.8E−01	8.8E−02	2.8E−01
Rollover	2.1E−02	1.2E−03	3.5E−02	1.3E−03	3.5E−02	1.3E−03	3.6E−02	1.8E−03	4.3E−02	2.9E−03	5.3E−02	3.2E−03	5.6E−02	4.7E−03	6.8E−02	4.9E−03	7.0E−02
Other	1.0E−01	3.6E−03	6.0E−02	8.4E−03	9.1E−02	1.2E−02	1.1E−01	1.5E−02	1.2E−01	1.7E−02	1.3E−01	1.5E−02	1.2E−01	1.4E−02	1.2E−01	1.6E−02	1.2E−01
Environmental conditions																	
Normal/clear	9.4E−01	1.0E−01	3.0E−01	1.0E−01	3.0E−01	1.1E−01	3.1E−01	1.2E−01	3.3E−01	1.3E−01	3.4E−01	1.2E−01	3.3E−01	1.3E−01	3.3E−01	1.3E−01	3.3E−01
Rain	5.5E−02	7.9E−03	8.9E−02	9.4E−03	9.6E−02	9.9E−03	9.9E−02	7.8E−03	8.8E−02	6.5E−03	8.0E−02	5.1E−03	7.1E−02	3.0E−03	5.4E−02	5.7E−03	7.5E−02
Fog	8.1E−05	8.4E−06	2.9E−03	1.1E−05	3.3E−03	1.4E−05	3.7E−03	8.4E−06	2.9E−03	2.5E−05	5.0E−03	8.4E−06	2.9E−03	0.0E+00	0.0E+00	5.6E−06	2.4E−03
Wind	9.0E−04	8.4E−05	9.2E−03	1.9E−04	1.4E−02	1.6E−04	1.3E−02	1.1E−04	1.1E−02	1.8E−04	1.3E−02	1.2E−04	1.1E−02	2.8E−05	5.3E−03	3.4E−05	5.8E−03
Good light	7.9E−01	4.2E−02	2.0E−01	5.5E−02	2.3E−01	6.8E−02	2.5E−01	1.2E−01	3.3E−01	1.5E−01	3.6E−01	1.0E−01	3.0E−01	1.1E−01	3.1E−01	1.4E−01	3.5E−01
Residential	5.9E−01	6.2E−02	2.4E−01	5.6E−02	2.3E−01	6.0E−02	2.4E−01	6.9E−02	2.5E−01	8.3E−02	2.8E−01	8.4E−02	2.8E−01	8.7E−02	2.8E−01	8.8E−02	2.8E−01
Commercial	3.5E−01	4.1E−02	2.0E−01	4.5E−02	2.1E−01	4.9E−02	2.1E−01	5.2E−02	2.2E−01	4.7E−02	2.1E−01	4.0E−02	2.0E−01	3.8E−02	1.9E−01	3.9E−02	1.9E−01
Industrial	5.1E−02	6.3E−03	7.9E−02	6.8E−03	8.2E−02	7.5E−03	8.6E−02	7.7E−03	8.7E−02	7.3E−03	8.5E−02	5.8E−03	7.6E−02	4.6E−03	6.8E−02	4.8E−03	6.9E−02
Macroeconomic indicator																	
GDP per cápita (US $)	6.8E+03	5.1E+03	*	6.3E+03	*	7.2E+03	*	7.9E+03	*	8.0E+03	*	7.9E+03	*	6.0E+03	*	5.8E+03	*
CPI (2010 = 100)	1.1E+02	9.8E+01	*	1.0E+02	*	1.0E+02	*	1.1E+02	*	1.1E+02	*	1.1E+02	*	1.2E+02	*	1.3E+02	*
Motorization Index[a]	6.0E+00	4.1E+00	*	5.6E+00	*	7.0E+00	*	6.8E+00	*	6.2E+00	*	6.9E+00	*	5.9E+00	*	5.2E+00	*

Note: Std. D. = standard deviation; - = missing observation. Notation: $\alpha E \pm \beta = \alpha \times 10^{\pm\beta}$.

[a]New vehicle sales per 1000 people.

*Yearly record (no standard deviation).

Marginal effects $ME(P_i, x_{ni})$ describe the average change in probability P_i from a unitary change in variable x_{ni} (Equation 5).

$$ME(P_i, x_{ni}) = \frac{\delta P_i}{\delta x_{ni}} \quad (5)$$

The results after applying this method on the described data set are presented and discussed next.

Estimation Results

This section presents and discusses the estimation results of the MNL model with random parameters for CIS considering PDO and IFO. The estimation process involved three stages: (*a*) MNL estimation with classic (Table 1) and available intuitive variables to be consistent with previous literature and the specific context of Medellín; (*b*) adding macroeconomic indicators to previous MNL to be confident that significance is related to these variables and not to omitted correlated variables; and (*c*) testing random parameters. After carefully testing several model structures, the MNL with random parameters that represents the best specification for CIS is presented in Table 3. It is estimated with the software NLOGIT. Computation time is 30.11 h.

The variables in the final model are significant and have intuitive signs (Table 3). Random parameters follow normal distributions. Table 3 describes the variables in the first column and shows the corresponding parameters, *t*-statistics, and ME in the subsequent columns. The mean value of the random parameters are presented over their standard deviations in parenthesis.

The likelihood ratio test statistic (Equation 6) assesses the overall significance of the MNL with random parameters (labeled U for unrestricted model), over its fixed-parameter version, that is, the regular MNL (labeled R for restricted model).

$$\chi^2 = -2\left[LL(\beta_R) - LL(\beta_U)\right] \quad (6)$$

After model estimation, the corresponding log-likelihood functions at convergence are $LL(\beta_U) = -12{,}9203.9$, and $LL(\beta_R) = -12{,}9932.9$. Therefore, $\chi^2 = 1{,}458$ and distributed with five degrees of freedom (five more parameters estimated in the mixed logit model, that is, standard deviations of random parameters). Therefore, with at least a 99.9% level of confidence, the MNL can be rejected and the MNL with random parameters is preferred.

The base scenario modeled in Table 3 is described below to better understand the impact of each variable in CIS, that is:

- Roadway geometry and traffic control: Crashes on roadways with three or more lanes, two or more carriageways, tangent segments, no intersections, no roundabout, no traffic light, and bidirectional/reversible/counter-flow traffic.
- Pavement conditions: Crashes on wet pavement or pavement with loose material.
- Temporal attributes: Crashes in December, during weekdays (Monday to Friday), between 10.00 a.m. and 6.00 p.m.
- Crash-specific attributes: Regular crashes without run overs, occupants falling, rollover, or other unusual circumstance.
- Environmental attributes: Crashes with no rain, no fog, no wind, with bad or nonexistent lighting, on areas with industrial/commercial land use.
- Macroeconomic indicators: All at zero values.
- Key findings associated with each group of variables are discussed next.

Roadway Geometry and Traffic Control

As the number of lanes around the accident decreases, the likelihood of suffering a more severe accident increases. On average, the probability of IFO increases by 1.02% on one lane. The parameter for two-lane roadways is random. On average, two-lane roads increase the probability of IFO by 1.2%. Random effects indicate 45.64% of the observations have IFO increments, and 54.36% IFO reductions, both at different scales (normally distributed), which suggests there is an interesting transition in CIS from one lane to three or more via two-lane roads. The increment in IFO might be related to on-street parking on urban two-lane roads. Notice that a couple of studies found the greater the number of lanes the higher the probability of CIS (*27, 28*), which partially aligns with the 45.64% of crashes with CIS increments. However, discrepant conclusions are attributable to their focus on pedestrian crashes because when pedestrians cross more lanes they are more exposed to the traffic and more vulnerable.

Single carriageway roadways tend to decrease the severity of the accidents; on average the probability of an IFO decreases by 1.25% under this condition. Driving speeds on single-carriageway roads are typically slow, which might mitigate the occurrence of severe accidents.

The parameter for crash at intersection is random. On average, an intersection increases the probability of IFO by 1.45%. Random effects indicate that 74.43% of the observations have IFO increments, and 25.57% IFO decrements, both at different scales (normally distributed). IFO increments could be attributed to many intersections with a larger propensity of rear-end collisions as well as angle or side impacts. Additionally, the combination of driver aggressiveness and traffic guidance ambiguity can increase IFO. The IFO reduction could be attributed to intersections with higher standards and clearer information for traffic management. This aspect has been confirmed in the context of developing countries by another study (*28*).

The presence of traffic signals on average decreases the probability of IFO by 2.90%. Since signals are placed to

Table 3. MNL Model with Random Parameters for CIS: Damage Only (Base Utility), and Injuries Including Fatalities

Variable description	Coeff.	*t*-stat	Avg. *ME* (%)
Constant specific to the injuries and fatalities	8.39E−01	3.06E+00	-
Roadway geometry and traffic control			
Indicator variable for crash on one-lane roadway	3.45E−01	1.54E+01	1.02E+00
Indicator variable for crash on two-lane roadway	−1.49E−01 (1.39E+00)	−6.71E+00 (2.25E+01)	1.16E+00 (2.78E+00)
Indicator variable for crash on single carriageway roadway	−1.64E−01	−9.87E+00	−1.25E+00
Indicator variable for crash on curve	1.10E−01	4.76E+00	1.06E+00
Indicator variable for crash on roundabout	−6.05E−01	−1.37E+01	−1.47E−01
Indicator variable for crash on intersection	5.85E−01 (8.91E−01)	2.00E+01 (5.69E+00)	1.45E+00 (3.61E+00)
Indicator variable for crash in the presence of traffic light	−3.77E−01	−1.39E+01	−2.90E+00
Indicator variable for crash on unidirectional roadway	−2.25E−01	−1.39E+01	−1.56E+00
Pavement conditions			
Indicator variable for crash on dry pavement	2.35E−01	7.01E+00	2.32E+00
Temporal attributes			
Indicator variable for crash in January	1.55E−01	4.76E+00	1.18E−01
Indicator variable for crash in February	1.20E−01	3.80E+00	1.02E−01
Indicator variable for crash in March	1.09E−01	3.54E+00	1.01E−01
Indicator variable for crash in April	1.24E−01	4.00E+00	7.03E−02
Indicator variable for crash in May	7.83E−02	2.56E+00	7.26E−02
Indicator variable for crash in June	6.85E−02	2.21E+00	5.97E−02
Indicator variable for crash in July	1.18E−01	3.77E+00	1.00E−01
Indicator variable for crash in August	5.10E−02	1.64E+00	4.40E−02
Indicator variable for crash in September	1.19E−01	3.89E+00	1.11E−01
Indicator variable for crash in October	8.76E−02	2.88E+00	8.24E−02
Indicator variable for crash in November	9.80E−02	3.18E+00	6.40E−03
Indicator variable for crash on Saturday	9.35E−02	5.18E+00	1.50E−01
Indicator variable for crash on Sunday	4.17E−01 (1.07E+00)	1.44E+01 (1.05E+01)	4.93E−01 (2.04E+00)
Indicator variable for crash between 06 PM and 07 PM	9.43E−02	3.26E+00	5.32E−02
Indicator variable for crash between 07 PM and 08 PM	1.63E−01	5.81E+00	1.02E−01
Indicator variable for crash between 08 PM and 09 PM	2.74E−01	9.25E+00	1.63E−01
Indicator variable for crash between 09 PM and 10 PM	5.60E−01	1.36E+01	1.13E−02
Indicator variable for crash between 10 PM and 11 PM	7.99E−01	1.82E+01	2.69E−01
Indicator variable for crash between 11 PM and 12 PM	9.56E−01	1.90E+01	2.57E−01
Indicator variable for crash between 12 PM and 1 AM	8.42E−01	1.09E+01	7.03E−02
Indicator variable for crash between 01 AM and 2 AM	9.75E−01	1.32E+01	9.48E−02
Indicator variable for crash between 02 AM and 3 AM	9.45E−01	1.20E+01	7.99E−02
Indicator variable for crash between 03 AM and 4 AM	9.56E−01	1.16E+01	6.32E−02
Indicator variable for crash between 04 AM and 5 AM	1.09E+00	1.29E+01	8.03E−02
Indicator variable for crash between 05 AM and 6 AM	1.34E+00	2.03E+01	2.17E−01
Indicator variable for crash between 06 AM and 7 AM	1.06E+00	2.16E+01	3.48E−01
Indicator variable for crash between 07 AM and 8 AM	7.55E−01	2.12E+01	4.73E−01
Indicator variable for crash between 08 AM and 9 AM	5.68E−01	1.77E+01	3.61E−01
Indicator variable for crash between 09 AM and 10 AM	1.99E−01	6.75E+00	1.11E−01
Crash-specific attributes			
Indicator variable for crash involving run over	1.28E+01	1.25E+01	4.30E−03
Indicator variable for crash involving falling out of vehicle	1.22E+01	1.20E+01	4.10E−03
Indicator variable for crash involving other characteristics	5.04E+00	3.37E+01	1.46E+00
Indicator variable for crash involving vehicle rollover	4.61E+00	3.21E+01	3.81E−01
Environmental attributes			
Indicator variable for crash with rain, fog, or wind	−3.92E−01 (8.89E−01)	−8.84E+00 (7.65E+00)	−1.12E−01 (6.62E−01)
Indicator variable for crash with good lighting	1.49E−01	7.15E+00	1.26E+00
Indicator variable for crash in residential area	5.16E−01	2.49E+01	3.08E+00

(continued)

Table 3. (continued)

Variable description	Coeff.	*t*-stat	Avg. *ME* (%)
Macroeconomic indicators			
Gross Domestic Product per capita (current thousand US$)	−9.45E−02 (1.30E−01)	−7.45E+00 (7.62E+00)	−4.85E+00 (3.47E+00)
Consumer price index standardized to 2010, i.e., CPI = 100	−3.14E−02	−8.11E+00	−3.64E+01
Motorization index (vehicles/1,000 people)	3.14E−02	1.35E+01	3.26E+01

Note: 306,576 observations. Log likelihood at convergence = −129,203.9. Random parameters normally distributed and estimated with 200 Halton draws. Standard deviations in parenthesis. Notation: $\alpha E \pm \beta = \alpha \times 10^{\pm\beta}$.

guide and control traffic, they minimize ambiguity for drivers and hence reduce conditions for IFO. This conclusion aligns with the results reported by Kamruzzamann et al. (*32*).

Crashes on roundabouts on average reduce the probability of suffering IFO by 0.147%. Proper traffic guidance, ability to merge vehicles at near-parallel trajectories, reduced speeds, and greater stopping sight distances might promote safer conditions and hence reduce IFO.

Crashes on unidirectional roads tend to be less severe. On average the probability of suffering an IFO decreases by 1.56%. Unidirectional roads remove the possibility of head-on collisions by traffic in opposite directions, which reduces the intensity of the crashes and hence their severity. Notice that this finding differs from the results obtained by Kamruzzamann et al. where two-way roads decreased CIS (*32*) and might be related to unobserved heterogeneity.

Pavement Conditions

Interestingly, dry pavement conditions tend to worsen CIS. On average, the probability of IFO increases by 2.32%. Dry pavements are expected to enhance vehicle performance and increase safety, which might reduce crash frequency. However, dry pavements might also generate a behavioral sense of comfort that reduces risk aversion (e.g., speeding, high acceleration/deceleration, aggressive cornering, among others), which increases severity. Notice that previous studies found reductions in CIS in wet pavement conditions for drivers under 45 years old (*26*), which aligns with the results from the model.

Temporal Attributes

Every other month has a higher probability of IFO than December (base-case). December involves leisure and festivities in Medellín. Traffic agencies intensify crash prevention and enforcement to reduce unsafe driving during this month. Likewise, December celebrations are more family oriented, which might promote positive attitudes and reduce risky behavior. On the other hand, January has the highest increase in the average probability of IFO (0.12% on average). Similar to December, January hosts leisure activities and festivities. There are two main holidays in January: the secular New Year's Day (January 1st), and the Catholic Epiphany—or Three Kings' Day—(January 6th). These festivities are slightly less family oriented, usually involve vacation traveling, and some users might drive under the influence of alcohol and other substances (DUI). Such an unfortunate combination might reinforce the corresponding increase in injury severity. Other months with high rates of IFO crashes are September (0.11% on average), February, and March (0.10% on average each), which might encapsulate unobserved heterogeneity associated with such periods.

Crashes are more severe on weekends than weekdays. The probability of suffering IFO increases by 0.15% on average on Saturdays. The parameter for Sunday is random. On average, Sunday increases the probability of IFO by 0.49%. Random effects indicate 65.16% of the observations have IFO increments, and 34.84% IFO reduction, both at different scales (normally distributed). Several attributes might make weekend crashes more severe. For example, DUI cases are expected to be higher after midnight on Friday and Saturday. The influence of alcohol and drugs can increase aggressive driving and risky behaviors, which increases CIS. Likewise, pedestrians, cyclists, and motorcyclists under the influence of alcohol or drugs can generate situations with intensified CIS. On the other hand, some Sundays might be less hectic and more leisure-oriented, which might explain IFO decrements. The work by Tulu et al. supports these conclusions (*28*).

The tendency to suffer IFO crashes by hour of day is shown in Figure 1. There is an increasing trend from 6:00 p.m. (0.05% on average) until midnight (0.26% on average), then it remains between 0.07 and 0.09% from midnight to 5:00 a.m. Next, it goes up again until 8:00 a.m with a 0.47% peak average IFO increment between 7:00 and 8:00 a.m. Afterward, the probability of IFO goes down to base conditions at 10:00 a.m. The small increment during the evening peak might be due to the combination of congestion and darkness. The abrupt reduction at 9:00 p.m. can be explained by the system reaching capacity due to the evening commute. Congestion and flexible arrival times reduce travel speed, and, therefore, CIS. Morning peak was found to be the most critical time of day for CIS. This might be explained by hectic behavior during the morning commute. Previous research also found CIS increments during the morning peak due to fatigue of truck drivers (*10, 29*). Research in developing countries has found an increment in CIS during the night (*27, 28, 32*). However, the threshold

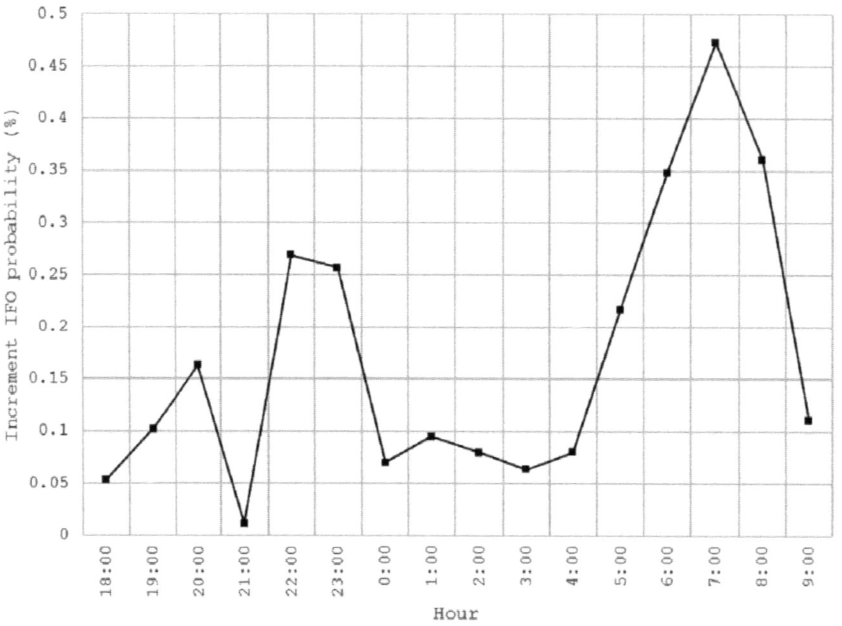

Figure 1. Estimated increment in IFO probability by hour of day (0% base case between 9:00 a.m. and 6:00 p.m.).

between night and day conditions changes from study to study, with 6:00 p.m. and 6:00 a.m. being the most common boundaries. This paper shows a more detailed description of the behavior of CIS by time of day.

Crash-Specific Attributes

Run overs increase the probability of IFO. On average, it goes up by the order of 0.01%. The nature of these events involves notorious damage to the person involved and, hence, increment in the probability of suffering IFO. Although Xie et al. (*11*) found run overs to critically deteriorate CIS with an ordered outcome approach, results from the MNL with random parameters estimated in this paper solve unobserved heterogeneity that might inflate such conclusions. Additionally, similar results for run overs have been found for pedestrians in another study (*32*).

Falling out of a vehicle increases the probability of IFO. On average, it goes up by the order of 0.01%. Since the occupant's body is exposed to extreme conditions, there is increased damage to the people involved. Motorcyclists and cyclists might account for many of these instances.

Vehicle rollover increases crash severity. On average, the probability of suffering IFO increases by 0.38%. Medellín is located in the Aburra Valley, a region with diverse topography, that is, level terrain at the center but more rolling/mountainous terrain at the edges. Rollovers may be more critical due to inclined hill sides. The usage of seatbelts is mandatory in the city, however, crashes may be more severe when occupants do not use them.

On average, crashes with other uncommon circumstances are 1.46% more likely to suffer IFO. Although details on such circumstances are not provided in the data set, this can be an avenue for additional investigations to understand them better.

Environmental Conditions

The parameter for the presence of rain, fog, and/or wind is random. On average, crashes under these conditions decrease the probability of IFO by 0.11%. Random effects indicate 32.96% of the observations have IFO increments, and 67.04% IFO decrements, both at different scales (normally distributed). Although the larger reduction in IFO seems counterintuitive, drivers might be aware of the hazards associated with such conditions and increase their level of attention, which might allow them to take precautionary action in avoiding the unfortunate case of a crash. Likewise, drivers will tend to reduce their speeds when driving in rain, fog, or wind. In the case of a crash, then, the intensity would be low as well as the corresponding severity. Notice that this finding mimics the intuition of dry pavements, presented above. On the other hand, the increment in IFO for lesser cases is telling that in some cases—perhaps with aggressive drivers—rain, fog, and/or wind increases IFO. Fog and rain reduce visibility. Rain also reduces friction. Wind can affect maneuverability. The combination of these factors and incautious users generate unfortunate conditions for severe crashes.

Good lighting increases the probability of severe crashes by 1.26% on average. Again, this counterintuitive result is similar to the case of dry pavement. Under good lighting, people might feel more comfortable and drive more aggressively. Therefore, when crashes occur, they tend to be more severe because of speeding, pedestrians and cyclists crossing streets in wrong places, etc.

Crashes in residential areas tend to be more severe than crashes in industrial/commercial areas. On average, the probability of an IFO increases by 3.08% in residential areas. Perhaps the presence of more vulnerable users on the streets in residential areas, and the fact that the urban form of Medellín does not always separate high-hierarchy roads from residential areas, might promote undesired interactions between fast vehicles and non-motorized users, with unfortunate consequences. Other studies found that crashes in residential areas increase the probability of injuries (*12*).

Macroeconomic Indicators

The parameter GDP per capita is random. On average, a 1,000 increment in GDP per capita reduces the probability of IFO by 4.85%. Random effects indicate 23.36% of the crashes have IFO increments, and 76.64% IFO reductions, both at different scales (normally distributed). Increments in the national GDP per capita approximate gains in productivity per person. Thus, as the people of Medellín become more productive, they tend to reduce their risky behavior. Perhaps this occurs because productive people can perceive higher income and expect a better economic status; all these expectations can be significantly affected by the experience of a severe crash. On the other hand, the increment in IFO might be associated with people who have risen suddenly to wealth or high positions. Inexperience and flamboyance might reduce driving carefulness and hence increase crash severity. The latter results align with Bishai (*33*), who found economic improvements increased CIS.

Increments in CPI tend to decrease the severity of accidents. On average, a 1% change in this index decreases the probability of IFO by 0.36%. Having household members involved in severe crashes can be extremely critical if the cost of living gets higher. People under such an economic environment are expected to be more careful in their decisions, which might result in reductions in the probability of experiencing severe crashes.

As the motorization index increases, the probability of severe crashes also increases. On average, a unitary increase in this ratio increases the probability of IFO by 0.33%. More vehicles on the streets might be associated with new, inexperienced users which will increase the chances of crashes and their corresponding severity.

Policy Implications

The previous results support the following recommendations to improve CIS.

- Prioritizing intersection upgrades with signals can decrease the probability of IFO, especially if properly coordinated with level-of-service and capacity improvements.
- Since some good roadway conditions (e.g., dry pavements, good lighting, and no rain/fog/wind) increase CIS, it is important to incorporate forgiving designs in new roads to improve mobility and mitigate CIS with validated products (e.g., longitudinal safety barriers, crash cushion, and pavement markings, among others).
- Prioritize activities of education, outreach, and control to reduce CIS during periods when it is worse, such as, mornings and nights, weekends, and January.
- Falling out of vehicle provides an approximation of the negative impact of motorcycle/bicycle crashes in CIS. Better roadway conditions should be provided for these users to mitigate their CIS, for example, motorcycle-/bicycle-only routes, proper segregation of mixed traffic, among others.
- Investigate the attributes aggravating CIS in residential areas, and develop urban planning policies to mitigate them.
- Investigate population segments susceptible to more severe crashes during periods of good economic/productivity conditions. Target traffic safety activities of education, outreach, and control to such populations.
- Develop policy and infrastructural readiness to mitigate CIS in periods with high motorization rates, which is found to deteriorate CIS.

The next section summarizes this CIS investigation and its key findings. It also discusses limitations and provides direction for future research efforts.

Conclusion

This paper investigates CIS oin an urban area in a developing country. The contributions of the paper are threefold: (*a*) providing an econometric framework to understand urban CIS in developing countries, (*b*) employing macroeconomic conditions as explanatory variables in this context, and (*c*) providing key findings to influence policy-making and improve CIS.

A MNL model with random parameters is estimated using a unique data set of crashes for Medellín, Colombia. In previous literature many variables have been found to be significant and intuitive to understand CIS in this region, related to roadway geometry, traffic control, pavement characteristics, temporal attributes, crash-specific characteristics, and environmental conditions. Additionally, macroeconomic indicators associated with GDP per capita, CPI, and motorization index are found to explain CIS while incorporating socioeconomic and idiosyncratic conditions unique to urban areas in developing countries like Medellín.

The top variables for CIS reductions include GDP per capita, traffic lights, unidirectional flows, and single carriageways. On the other hand, top variables for CIS increments include residential areas, dry pavements, other uncommon crashes, and intersections. Variables associated with random effects (increments/decrements) include two-lane roadways, intersections, Sunday, the presence of rain, fog or wind; and GDP per capita.

Most of the variables triggering IFO crashes are intuitive. Remarkably, however, some variables, like dry pavements, good lighting, and no rain, fog or wind, tend to increase CIS. These counterintuitive results indicate users might reduce their thresholds of risk when traveling on roads with good conditions, which might worsen severity when crashes occur.

Other classic variables tend to produce very intuitive outcomes associated with CIS. The new variables for macroeconomic conditions indicate that crashes in urban areas in developing countries might be less severe when economic productivity rises (higher GDP per capita) but in some few cases may be more severe. Additionally, increments in CPI reduce CIS because higher cost of living might increase the risk awareness of transportation users. Furthermore, motorization increases CIS, as well as GDP per capita for a reduced number of observations. This might be related to new, inexperienced drivers (motorization) and more flamboyant users (GDP per capita).

The results support several policies that can mitigate CIS in Medellín. Some examples include: upgrades to signalization at intersections; incorporation of forgiving roadway designs; development of better conditions for motorcyclists and nonmotorized users; and prioritization of education, outreach, and control campaigns during periods of good macroeconomic conditions (for some segments of the population), high motorization rates, and other specific times, like mornings, nights, peak hours, weekends, Januaries, etc.

Several future extensions can enhance the findings from this study. More research can be developed to understand the underrepresentation of fatalities in the data set, one potential direction might be to expand the geographic scope of the data set. Additionally, it would be interesting to find ways to expand the records for fatalities, for example, by developing upsampling procedures to simulate missing observations. Although demographic and crash-specific information is available for vehicles and people that suffered injuries and fatalities, no additional data is available for PDO. It is suggested that such data be collected to improve the current analysis. Furthermore, it would be interesting to test the geographic transferability of the model against models in developed countries to provide a quantitative comparison of such different contexts. Temporal transferability might be also interesting to understand whether CIS trends are sustained over time or change in different years. Finally, it would be valuable to test the effect of different transportation policies implemented in Medellín in this time period to understand better their impacts on CIS.

Acknowledgments

The authors would like to thank the Medellín Mobility Agency (Secretaría de Movilidad de Medellín) for sharing the crash database; the Group for Transportation Research and Cross-Disciplinary Studies (TREX) at the Florida Institute of Technology for intellectual guidance and computational resources; and the Center for Road Safety (CRS) at Purdue University for technical support.

References

1. World Health Organization. *Global Status Report on Road Safety 2015*. WHO Library Cataloguing-in-Publication Data Global, 2015, p. 340. https://doi.org/10.1136/injuryprev-2013-040775.
2. Moreno Lozada, S. L. *Datos Para La Vida: Forensis 2016*. Bogota, Colombia, 2017.
3. World Health Organization. *Global Status Report on Road Safety 2015*. WHO Library Cataloguing-in-Publication Data Global, 2015, p. 340. https://doi.org/10.1136/injuryprev-2013-040775.
4. Bogue, S., R. Paleti, and L. Balan. A Modified Rank Ordered Logit Model to Analyze Injury Severity of Occupants in Multivehicle Crashes. *Analytic Methods in Accident Research*, Vol. 14, 2017, pp. 22–40. https://doi.org/10.1016/j.amar.2017.03.001.
5. Yasmin, S., N. Eluru, A. R. Pinjari, and R. Tay. Examining Driver Injury Severity in Two Vehicle Crashes: A Copula Based Approach. *Accident Analysis & Prevention*, Vol. 66, 2014, pp. 120–135. https://doi.org/10.1016/j.aap.2014.01.018.
6. Kaplan, S., and C. G. Prato. Risk Factors Associated with Bus Accident Severity in the United States: A Generalized Ordered Logit Model. *Journal of Safety Research*, Vol. 43, No. 3, 2012, pp. 171–180. https://doi.org/10.1016/j.jsr.2012.05.003.
7. Weiss, H. B., S. Kaplan, and C. G. Prato. Analysis of Factors Associated with Injury Severity in Crashes Involving Young New Zealand Drivers. *Accident Analysis & Prevention*, Vol. 65, 2014, pp. 142–155. https://doi.org/10.1016/j.aap.2013.12.020.
8. Feng, S., Z. Li, Y. Ci, and G. Zhang. Risk Factors Affecting Fatal Bus Accident Severity: Their Impact on Different Types of Bus Drivers. *Accident Analysis & Prevention*, Vol. 86, 2016, pp. 29–39. https://doi.org/10.1016/j.aap.2015.09.025.
9. Yasmin, S., N. Eluru, C. R. Bhat, and R. Tay. A Latent Segmentation Based Generalized Ordered Logit Model to Examine Factors Influencing Driver Injury Severity. *Analytic Methods in Accident Research*, Vol. 1, 2014, pp. 23–38. https://doi.org/10.1016/j.amar.2013.10.002.
10. Hao, W., C. Kamga, and D. Wan. The Effect of Time of Day on Driver's Injury Severity at Highway-Rail Grade Crossings in the United States. *Journal of Traffic and Transportation Engineering (English Edition)*, Vol. 3, No. 1, 2016, pp. 37–50. https://doi.org/10.1016/j.jtte.2015.10.006.
11. Xie, Y., Y. Zhang, and F. Liang. Crash Injury Severity Analysis Using Bayesian Ordered Probit Models. *Journal of Transportation Engineering*, Vol. 135, No. 1, 2009, pp. 18–25. https://doi.org/10.1061/(ASCE)0733-947X(2009)135:1(18).
12. Bandyopadhyaya, R., and S. Mitra. Modelling Severity Level in Multi-Vehicle Collision on Indian Highways. *Procedia: Social and Behavioral Sciences*, Vol. 104, 2013, pp. 1011–1019. https://doi.org/10.1016/j.sbspro.2013.11.196.

13. Zeng, Z., W. Zhu, R. Ke, J. Ash, Y. Wang, J. Xu, and X. Xu. A Generalized Nonlinear Model-Based Mixed Multinomial Logit Approach for Crash Data Analysis. *Accident Analysis & Prevention*, Vol. 99, 2017, pp. 51–65. https://doi.org/10.1016/j.aap.2016.11.008.
14. Gong, L., and W. D. Fan. Modeling Single-Vehicle Run-off-Road Crash Severity in Rural Areas: Accounting for Unobserved Heterogeneity and Age Difference. *Accident Analysis & Prevention*, Vol. 101, 2017, pp. 124–134. https://doi.org/10.1016/j.aap.2017.02.014.
15. Cerwick, D. M., K. Gkritza, M. S. Shaheed, and Z. Hans. A Comparison of the Mixed Logit and Latent Class Methods for Crash Severity Analysis. *Analytic Methods in Accident Research*, Vol. 3–4, 2014, pp. 11–27. https://doi.org/10.1016/j.amar.2014.09.002.
16. Behnood, A., and F. L. Mannering. An Empirical Assessment of the Effects of Economic Recessions on Pedestrian-Injury Crashes Using Mixed and Latent-Class Models. *Analytic Methods in Accident Research*, Vol. 12, 2016, pp. 1–17. https://doi.org/10.1016/j.amar.2016.07.002.
17. Behnood, A., and F. L. Mannering. The Temporal Stability of Factors Affecting Driver-Injury Severities in Single-Vehicle Crashes: Some Empirical Evidence. *Analytic Methods in Accident Research*, Vol. 8, 2015, pp. 7–32. https://doi.org/10.1016/j.amar.2015.08.001.
18. Behnood, A., A. M. Roshandeh, and F. L. Mannering. Latent Class Analysis of the Effects of Age, Gender, and Alcohol Consumption on Driver-Injury Severities. *Analytic Methods in Accident Research*, Vol. 3–4, 2014, pp. 56–91. https://doi.org/10.1016/j.amar.2014.10.001.
19. Chen, C., G. Zhang, Z. Tian, S. M. Bogus, and Y. Yang. Hierarchical Bayesian Random Intercept Model-Based Cross-Level Interaction Decomposition for Truck Driver Injury Severity Investigations. *Accident Analysis & Prevention*, Vol. 85, 2015, pp. 186–198. https://doi.org/10.1016/j.aap.2015.09.005.
20. Shaheed, M. S., and K. Gkritza. A Latent Class Analysis of Single-Vehicle Motorcycle Crash Severity Outcomes. *Analytic Methods in Accident Research*, Vol. 2, 2014, pp. 30–38. https://doi.org/10.1016/j.amar.2014.03.002.
21. Xie, Y., K. Zhao, and N. Huynh. Analysis of Driver Injury Severity in Rural Single-Vehicle Crashes. *Accident Analysis & Prevention*, Vol. 47, 2012, pp. 36–44. https://doi.org/10.1016/j.aap.2011.12.012.
22. Islam, S., and A. B. Hossain. Comparative Analysis of Injury Severity Resulting from Pedestrian–Motor Vehicle and Bicycle–Motor Vehicle Crashes on Roadways in Alabama. *Transportation Research Record: Journal of the Transportation Research Board*, 2015. 2514: 79–87.
23. Shaheed, M. S. B., K. Gkritza, W. Zhang, and Z. Hans. A Mixed Logit Analysis of Two-Vehicle Crash Severities Involving a Motorcycle. *Accident Analysis & Prevention*, Vol. 61, 2013, pp. 119–128. https://doi.org/10.1016/j.aap.2013.05.028.
24. Yasmin, S., and N. Eluru. Evaluating Alternate Discrete Outcome Frameworks for Modeling Crash Injury Severity. *Accident Analysis & Prevention*, Vol. 59, 2013, pp. 506–521. https://doi.org/10.1016/j.aap.2013.06.040.
25. Wu, Q., G. Zhang, Y. Ci, L. Wu, R. A. Tarefder, and A. "Dely" Alcántara. Exploratory Multinomial Logit Model-Based Driver Injury Severity Analyses for Teenage and Adult Drivers in Intersection-Related Crashes. *Traffic Injury Prevention*, Vol. 17, No. 4, 2016, pp. 413–422. https://doi.org/10.1080/15389588.2015.1100722.
26. Morgan, A., and F. L. Mannering. The Effects of Road-Surface Conditions, Age, and Gender on Driver-Injury Severities. *Accident Analysis & Prevention*, Vol. 43, No. 5, 2011, pp. 1852–1863. https://doi.org/10.1016/j.aap.2011.04.024.
27. Verzosa, N., and R. Miles. Severity of Road Crashes Involving Pedestrians in Metro Manila, Philippines. *Accident Analysis & Prevention*, Vol. 94, 2016, pp. 216–226. https://doi.org/10.1016/j.aap.2016.06.006.
28. Tulu, G. S., S. Washington, M. M. Haque, and M. J. King. Injury Severity of Pedestrians Involved in Road Traffic Crashes in Addis Ababa, Ethiopia. *Journal of Transportation Safety & Security*, Vol. 9, No. 1, 2017, pp. 47–66. https://doi.org/10.1080/19439962.2016.1199622.
29. Pahukula, J., S. Hernandez, and A. Unnikrishnan. A Time of Day Analysis of Crashes Involving Large Trucks in Urban Areas. *Accident Analysis & Prevention*, Vol. 75, 2015, pp. 155–163. https://doi.org/10.1016/j.aap.2014.11.021.
30. Wei, F., Z. Wang, P. S. Lin, P. P. Hsu, S. Ozkul, J. Jackman, and M. Bato. Safety Effects of Street Illuminance at Urban Signalized Intersections in Florida. *Transportation Research Record: Journal of the Transportation Research Board*, 2016. 2555: 95–102.
31. Altwaijri, S., M. Quddus, and A. Bristow. Analysing the Severity and Frequency of Traffic Crashes in Riyadh City Using Statistical Models. *International Journal of Transportation Science and Technology*, Vol. 1, No. 4, 2012, pp. 351–364. https://doi.org/10.1260/2046-0430.1.4.351.
32. Kamruzzaman, M., M. Haque, and S. Washington. Analysis of Traffic Injury Severity in Dhaka, Bangladesh. *Transportation Research Record: Journal of the Transportation Research Board*, 2014. 2451: 121–130.
33. Bishai, D., A. Quresh, P. James, and A. Ghaffar. National Road Casualties and Economic Development. *Health Economics*, Vol. 15, No. 1, 2006, pp. 65–81. https://doi.org/10.1002/hec.1020.
34. World Development Indicators | DataBank. http://databank.worldbank.org/data/reports.aspx?source=world-development-indicators. Accessed July 31, 2017.
35. Train, K. E. *Discrete Choice Methods with Simulation*. Working Paper. University of California Berkley, Department of Economics, 1999.
36. Bhat, C. R. Simulation Estimation of Mixed Discrete Choice Models Using Randomized and Scrambled Halton Sequences. *Transportation Research Part B: Methodological*, Vol. 37, No. 9, 2003, pp. 837–855.
37. Train, K. *Halton Sequences for Mixed Logit*. Working Papers, 1999, pp. 1–18.

The Standing Committee on Transportation in the Developing Countries (ABE90) peer-reviewed this paper (18-04655).

Article

TRR
JOURNAL OF THE TRANSPORTATION RESEARCH BOARD

Transportation Research Record
2018, Vol. 2672(31) 54–62
© National Academy of Sciences:
Transportation Research Board 2018
Article reuse guidelines:
sagepub.com/journals-permissions
DOI: 10.1177/0361198118758639
journals.sagepub.com/home/trr
SAGE

Macroscopic Spatial Analysis of Pedestrian Crashes in Medellin, Colombia

Cristhian Lizarazo[1] and Víctor Valencia[2]

Abstract

Pedestrians are regarded as vulnerable road users as they experience the most severe consequences in collisions with motor vehicles. Pedestrians were involved in fewer than 4% of all traffic accidents but represented more than 12% of traffic fatalities in Medellin, Colombia, from 2009 to 2016. A better understanding of the factors affecting the likelihood of such accidents is an area of interest across local and governmental agencies in Colombia. The paper provides a spatial lag model estimated using spatial two–stage least squares (S2SLS) method to analyze the impact of land use, socioeconomic factors, and transportation modes on vehicle–pedestrian collisions ("crashes"). Consideration is given to spatial spillovers, avoiding inconsistent and inefficient estimates. Commercial land use has a positive correlation with the number of crashes while residential land use correlates with fewer crashes. Analysis of transportation modes demonstrated a higher risk of crashes associated with public transit. In Medellin, buses stop at the passenger's request instead of at regular bus stops, which increases the likelihood of pedestrians being involved in accidents. The effect of taxi as transportation mode, however, analysed using the number of taxi stands as proxy, is that the number of crashes is reduced. In this transportation mode, the entire trip is covered from point of origin to destination, reducing exposure to the risks of walking. These results might provide a better understanding of the dynamics involved in pedestrian crashes while validating the importance of including spatial econometrics for safety analysis.

Motor vehicles are the primary cause of fatal unintentional injuries in Colombia. According to the Colombian Transportation Agency, collisions involving either pedestrians or motorcycles were among the most dangerous in terms of the number of fatalities from 2004 to 2014 (*1*). Pedestrians are regarded as vulnerable road users as they experience the most severe consequences in collisions with motor vehicles. In Medellin, from 2009 to 2016, pedestrians were involved in fewer than 4% of traffic accidents but represented more than 12% of traffic fatalities. These statistics are constantly increasing due to growth in traffic and population, especially in urban areas in Colombia.

As discussed by Lord and Mannering (*2*), a better understanding of the factors affecting the likelihood of vehicle–pedestrian collision ("crash") has been an area of interest among transportation engineers for many decades. Usually, this problem is tackled by observing the number of crashes occurring in a geographical space over a specified time of period. That study refers to some methodological issues regarding temporal and spatial correlation across observations. As the same authors note, some spatial correlation might be present.

This study analyzed the spatial distribution of vehicle–pedestrian collisions in the urban area of Medellin, Colombia. As established by (*3*), such collisions are not distributed homogeneously in a metropolitan area; some neighborhoods have higher rates of pedestrian causalities than others. Social and physical factors affecting pedestrian crashes are explored in this study, which looks for countermeasures and policy measures that help to reduce crash rates in Medellin. Specifically, the current paper seeks to identify how transportation modes in Medellin influence the likelihood of a user being involved in a crash. Different proxy variables are analyzed to understand better the effect of the transportation modes. The results offer better support to transportation decision making to improve pedestrian safety.

The remainder of this paper is organized as follows. The next section provides a review of related studies in the literature indicating factors affecting the likelihood of pedestrian crashes. The "Research Approach" section describes the data applied for modeling estimation and explains the modeling specification strategy and reasoning to account for spatial dependence in the project. The "Results" section discusses the results and the effects of each of the variables on the likelihood of pedestrian crashes. The final section presents the general conclusions of the study and future research directions.

Previous Work

The studies in spatial crash safety analysis can be divided into two broad categories. The first category includes research at a

[1]Center for Road Safety, Lyles School of Civil Engineering, Purdue University, West Lafayette, IN
[2]Universidad Nacional de Colombia, Medellín, Antioquia, Colombia

Corresponding Author:
Address correspondence to Cristhian Lizarazo: clizaraz@purdue.edu

disaggregated level analyzing specific locations with high crash statistics. The second group of studies assesses aggregation in the crash data preparation. In this category, safety studies examine the number of crashes located in a specific region by measures such as census tracks, census blocks, and traffic analysis zones (*4*). The scope of the analysis defines this second category as the primary interest of this paper.

The literature reports multiple area–level multivariate models characterizing the factors affecting pedestrian crash risk. These models provide a robust characterization of variables influencing pedestrian collisions compared with those including only the exposure component (*5*). In general, factors affecting crash risk can be grouped into five categories: exposure or volume, land use characteristics, highway design, socioeconomic variables, and transportation-related variables.

Volume is an important component in vehicle–pedestrian collisions (*3, 5–10*). Based on (*9*), the number of vehicles entering a particular intersection is the most influential factor for frequency of pedestrian crashes. A 30% reduction in the traffic volume leads to a 35% reduction in the total number of injured pedestrians. Other studies have included not only vehicle volumes but also pedestrian volumes with manual and automated techniques (*10*). These studies are limited by the efforts required for extensive data collection to obtain data on pedestrian exposure.

A second important component in the analyzed literature explores the influence of land use on pedestrian collisions (*3, 5, 11, 12*). Based on these studies, commercial land use has a strong influence on increasing pedestrian collisions (*3, 12*). Other uses that tend to increase the likelihood of pedestrian collision are mixed use and parking lots. According to (*13*), industrial, commercial, and open land use types have greater likelihood for crashes while census tracts with a higher residential area have a significantly lower likelihood of pedestrian crashes. A greater number of schools also increases the likelihood of pedestrian collisions.

Highway design has been broadly discussed in the literature on designing safer environments for pedestrians. As the results show, intersections are the most dangerous for pedestrian collisions (*4*). Exclusive right turn lanes and wider roads tend to increase the likelihood of pedestrian collisions at intersections. One-way streets are more difficult environments for pedestrians due to lack of raised medians and refuge islands (*14*). This is supported by (*10*) which found an association of raised median with lower number of crashes. In addition, three-leg intersections are safer than four-leg intersections for pedestrians (*14*).

Analysis of socioeconomic characteristics includes a diverse number of factors to better identify vulnerable populations or places. It allows the adoption of better, region-specific policies and countermeasures for pedestrian safety (*5, 15*). In general, low-income and young populations, as well as Hispanic and less-educated people, are more vulnerable to being involved in crashes (*16–18*). Other models have been developed for specific population groups, such as finding physical variables for children and school zones, which found similar effects of the analyzed factors (*19, 20*).

An additional category of interest is transportation–related variables. In this case, the literature provides a positive relationship of crashes and number of weekly trips. Characterization of transportation modes reveals a greater number of crashes when more public transit is present in the area. In this case a proxy of transit is assessed through factors such as number of subway stations and bus stops (*14*). The study in (*20*) also supports a reduction in the number of crashes in taxi commuting in England.

The literature is consistent in determining the effects of the different factors on crash risk. However, in area crash modeling there are elements at one location which not only affect the risk at the specific location but also have a spatial spillover in nearby areas. By ignoring these spatial dependencies, a nonlinear model might produce inconsistent and inefficient estimates (*2, 21*). Similar discussions are provided in (*4, 15*) with additional studies emphasizing the necessity to account for spatial dependence in their models (*5, 9*).

The current study provides a spatial regression model to understand better how physical and social characteristics in Medellin affect vehicle–pedestrian collisions. To the best of the authors' knowledge, it is the first attempt in Colombia to identify factors influencing vehicle–pedestrian collisions. Lack of data in developing countries means a lack of reliable evidence for researchers to assess extensive collision models. A comparison with existing studies in developed countries might provide an insight into differences in terms of safety between Colombia and other regions. In addition, this information is valuable for multiple stakeholders including local agencies, insurance companies, and other companies in order to reduce risk. It provides a better understanding of how to implement safety changes across regions and adopt better regional policies to improve pedestrian safety.

Research Approach

Data Description

The study analyzes crashes that occurred in Medellin, Colombia, in the period of analysis, 2009 to 2016. The crashes are aggregated per Traffic Analysis Zones (TAZs) as defined by Steer Davis Gleave (SDG) in 2011 (*22*). Each TAZ is a spatial aggregation characterized as being homogeneous in terms of land use and population characteristics. Homogeneous variables in the model include number of residential units, socioeconomic status (a scale of 1–6 applied in Colombia for economic classification of neighborhoods) and income. Based on this information, TAZs seem to be a reasonable level of aggregation for the project. In addition, they are considered as a standard level of aggregation for crash analysis in the literature (*4, 15*). Only TAZs defined as urban were selected for analysis due to the higher incidence of

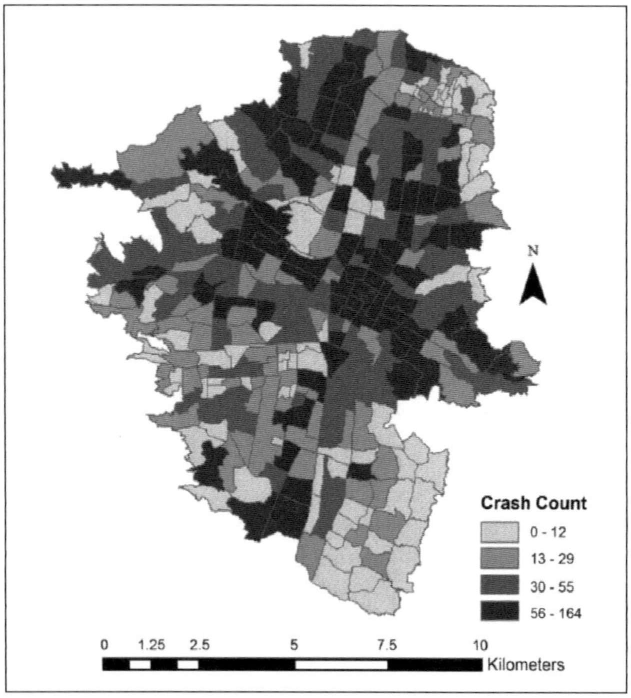

Figure 1. Crash count showing distribution of crashes in Medellin, Colombia.

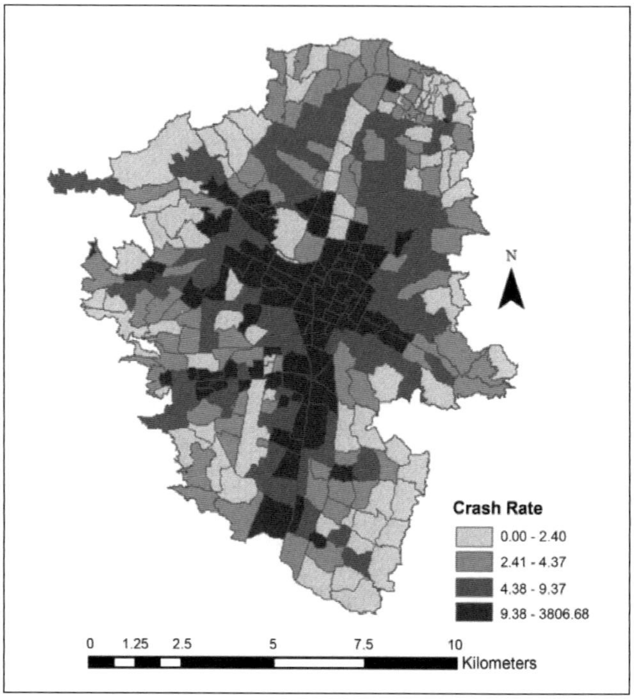

Figure 2. Crash rate showing distribution of crashes in Medellin, Colombia.

vehicle–pedestrian collisions in these areas. A total of 326 urban TAZs were examined in the project.

The crash data for analysis was obtained from the Medellin Mobility Department. A total of 13,075 vehicle–pedestrian collisions were reported in the region of interest from 2009 to 2016. They represented 12% of all the fatal crashes in that period. Geocoding was used to identify the location of the crashes. Incomplete data and typographical errors while registering the crashes did not permit location of every one; after geocoding it was possible to retrieve the location of 12,273 crashes. Figure 1 shows that most of the crashes occurred close to downtown and densely populated areas of Medellin where exposure is higher due to greater numbers of pedestrians.

A critical part of risk assessment is identifying how to represent exposure to the risk involved (*23*). In general, vehicle–pedestrian collisions are more likely to happen where there are pedestrian activities, as shown in Figure 1. To account for exposure, the number of pedestrian crashes is normalized by considering the population per TAZ. The new dependent variable, in this case, evaluates the total number of pedestrian crashes per 1,000 of population within the TAZ. After adjusting for exposure, the quantile map is shown in Figure 2. The adjusted crash rates are still concentrated in populated areas close to downtown areas but with a reduced effect compared with Figure 1.

Demographic and socioeconomic data were obtained from the 2013 transportation plan implemented in Medellin, Colombia. For each TAZ, the data included socioeconomic status, land use, and population. Infrastructure inventory was included in the analysis along with information on transportation modes to estimate their effect on crash risk. An inventory of the variables and descriptive statistics is shown in Table 1. Population–related variables were removed from the set of variables in the study to avoid endogeneity, considering the exposure adjustment.

Specification Strategy

Multiple approaches have been reported in the literature for modeling crash frequency–related data. When mapping crash locations to spatial units of analysis, it is important to consider spatial dependence to better estimate the effects of factors, as supported by (*15*). There are various specification search strategies in spatial econometrics modeling. The classic one is a sequential procedure sometimes referred to as specific-to-general. In this method, an initial model without spatially lagged variables is estimated. It is then tested for spatial autoregressive process and, if the null hypothesis of no spatial autocorrelation is rejected, remedial measures are applied (*24*). The alternative approach was proposed by Hendry (*25*); it starts with a very general model that is overparameterized in the sense that spatial correlation among various variables is assumed a priori. In this case, the Hendry strategy is a backward stepwise regression approach. It starts from the spatial common factor model, which is reduced on the basis of significant tests.

Table 1. Variables and Descriptive Statistics per Category

Variable	Definition	Mean	SD	Minimum	Maximum
Response variable					
crash_ra	Number of pedestrian crashes per 1,000 population	81.704	430.989	0	3806.685
Sociodemographic variables					
mal_per	Proportion of males	0.455	0.078	0	0.488
fem_per	Proportion of females	0.517	0.088	0	0.560
soc_stat	Socioeconomic status	2.972	1.548	0	6.000
Land use variables					
resid_area	Residential area in km^2	0.242	0.710	0	12.397
com_area	Commercial area in km^2	0.016	0.030	0	0.314
serv_area	Services area in km^2	0.027	0.035	0	0.216
greeni	Green area indicator	0.702	0.458	0	1.000
Classification of roads					
local	Proportion of local roads	0.582	0.252	0	1.000
collector	Proportion of collector roads	0.228	0.171	0	0.821
minor_art	Proportion of minor arterial roads	0.030	0.083	0	0.856
major_art	Proportion of major arterial roads	0.142	0.193	0	1.000
freeway	Proportion of freeways	0.017	0.062	0	0.492
Infrastructure variables					
length_sid	Length of sidewalks in km	4.234	2.799	0	15.516
road_le	Length of roadways in km	5.254	3.314	0.111	21.127
side_width	Average sidewalk width in m	1.334	0.774	0	6.133
Grade	Average road grades	5.285	4.186	0	17.214
Psig	Proportion signalized intersections	0.090	0.111	0	0.667
Bike	Length of bike lanes in km	0.839	1.017	0	5.354
Transportation mode					
new_transi	Length of transit routes in km	0.807	0.718	0	3.445
taxi_park	Number of taxi stands in the area	0.399	0.999	0	8.000
parkzi	Indicator parking garage inside the TAZ	0.064	0.246	0	1.000

Note: SD = standard deviation.

Based on the simulations conducted in (*24*), the classic approach seems to perform better compared with Hendry's approach. This result, along with the computational burden of applying Hendry's approach, supports the application of the classic strategy in the current study. The classic approach proposes as a first step estimating an initial model without spatially lagged variables which follows a standard set of regularity conditions such as an ordinary least squares regression (OLS), given by the expression

$$y = X\beta + \varepsilon \quad (1)$$

where y is the dependent variable, X is the vector of explanatory variables, ε is the error vector and β the model parameters. It is then tested whether substantive spatial dependence is present due to an omitted spatial lag or spatially autoregressive errors. These types of misspecifications are likely to show up if the true data generating processes are spatial lag model characterized by the expression

$$y = \rho Wy + X\beta + \mu \quad (2)$$

or the spatial AR error model given by

$$y = X\beta + (I - \lambda W)^{-1}\varepsilon \quad (3)$$

where W is a row–standardized matrix characterizing the spatial morphology and ρ and λ are scalar autoregressive parameters. Lagrange multiplier (LM) tests are applied to test substantive spatial dependence and its cause. The next section provides the results of the cited specification strategy.

Spatial Weights

Spatial weights are a key component in any cross-sectional analysis of spatial dependence. The weight expresses the neighbor structure between observations as an *nxn* matrix (*26*). The construction of the weight matrix can be assessed by following different procedures. The queen criterion is selected for structuring the weight matrix of the model, due to the irregular shape of the TAZs. In this method, neighbors are defined when two TAZs share a common edge or vertex.

Table 2. Ordinary Least Squares Model of Pedestrian–Vehicle Collisions in Medellin, Colombia

Variable	Parameter Estimate	Standard error	t-ratio	p-value
Constant	218.909	63.206	3.463	>.001
Sociodemographic variables				
Socioeconomic status	−71.132	13.634	−5.217	>.001
Land use				
Residential area in km^2	−191.612	35.184	−5.446	>.001
Commercial area in km^2	6705.273	904.188	7.416	>.001
Classification of roads				
Proportion of collector roads	184.589	124.893	1.478	.1404
Transportation mode				
Length of transit routes in km	79.778	31.092	2.566	.0107
Number of taxi stands	−115.314	45.980	−2.508	.0126
Indicator parking garages	192.689	85.811	2.245	.0254
R^2 = 0.2840		Adjusted R^2 = 0.2659		
Multicollinearity number 7.757		Jarque–Bera 16128.8 (>0.00001)		
Breusch–Pagan test 987.595 (0.00001)		Koenker–Bassett test 56.196 (0.00001)		

Results

Ordinary Least Squares Regression

The first step of the classic approach is to estimate an initial model without spatially lagged variables that follows a standard set of regularity conditions such as an OLS. The selection of variables was assessed through stepwise selection in SAS 9.4 based on multiple F-statistics. Highly correlated relations were avoided in the model.

The results of the OLS estimator are provided in Table 2. All the signs of the variables considered as significant are in line with the expectations. First, the common non-spatial diagnostics including possible indications of multicollinearity or heteroscedasticity are assessed. Multicollinearity was evaluated using the multicollinearity condition number. According to (*26*), the multicollinearity condition number is an indicator of the degree to which the columns in the X matrix show a linear relationship. As a rule of thumb, it is considered to be excessive when its value exceeds 30 to 50. The multicollinearity value equal to 7.75 did not exceed these thresholds and was considered acceptable. This low value was obtained since high correlated relations were removed from the data.

An additional diagnostic for OLS is the extent to which the regression error terms tend to follow normal distribution. The Jarque–Bera test relying on the third and fourth moments of the regression residuals is applied to evaluate normality. The value indicates that the errors are not normally distributed. The skew nature of the data might produce this result.

The final test evaluates the assumption of homoscedasticity. The Breusch–Pagan test and the Koenker–Bassett test are considered to this end. Based on the results, the null hypothesis of homoscedasticity is rejected. The difference between the Breusch–Pagan test and the Koenker–Bassett test supports the violation of normality as well.

Diagnostics for Spatial Dependence

Diagnostics to determine the presence of spatial dependence include Moran's I and LM. Moran's I is an index of spatial autocorrelation and typically varies (although it is not restricted) between −1 and 1, with 0 representing "complete spatial randomness." Moran's I is one of the most commonly applied test statistics for spatial autocorrelation. This statistic is characterized by the expression (*26*)

$$I = \frac{e'We / S_0}{e'e / n} \tag{4}$$

where e is the vector of the OLS residuals, W is the spatial weights matrix, n is the number of observational units and $S_0 = \sum_i \sum_j w_{ij}$ is the sum of the spatial weights. Based on the results, the Moran's I statistic is equal to 0.34. The test under randomization provided a p-value equal to 2.2e to 13, indicating misspecifications in the model.

Additional diagnostics for spatial dependence include the LM test statistics. LMs for the spatial lag model test the null hypothesis of $H_0 : \rho = 0$. The statistic is expressed as the ratio

$$\text{LM}_\rho = \frac{d_p^2}{D} \tag{5}$$

with the numerator term equal to

$$d_p = \frac{e'Wy}{\sigma^2_{\text{ML}}} \tag{6}$$

with Wy being the spatial lag term $\sigma^2_{\text{ML}} = e'e / n$ and the other terms as previously defined. The denominator can be expressed as

$$D = \left(WX\hat{\beta}\right)' \left[I - X\left(X'X\right)^{-1} X'\right]\left(WX\hat{\beta}\right) / \sigma^2_{\text{ML}} + T \tag{7}$$

Table 3. Lagrange Multiplier Test Diagnostics for Spatial Dependence

Test	Value	p-value
LM – Spatial error	27.634	1.466e−7
LM – Spatial Lag	38.356	5.895e−10
Robust LM – Spatial error	0.862	0.353
Robust LM – Spatial Lag	11.584	6.65e−4

Note: LM = Lagrange multiplier.

Equation 7 has two different components. The first part is the sum of the squared residuals of the spatially lagged predicted values ($WX\hat{\beta}$) on the X. The second term is a trace expression characterized by the expression

$$T = \mathrm{tr}\left(WW + W'W\right) \tag{8}$$

The statistic is distributed as χ^2 with one degree of freedom. The spatial error model has a similar structure estimated by

$$\mathrm{LM}_\lambda = \frac{d_\lambda^2}{T} \tag{9}$$

The numerator is

$$d_\lambda = \frac{e'We}{\sigma^2_{\mathrm{ML}}} \tag{10}$$

where We is a spatial lag. The LM tests were conducted to determine the nature of the spatial dependence. A robust version of these tests was proposed by (*27*) to correct for potential influence of other alternative specifications by subtracting a value. The robust test for the spatial lag test is given by the expression

$$\mathrm{LM}_\rho = \frac{\left(d_\rho - d_\lambda\right)^2}{\left(D - T\right)} \sim \chi^2(1) \tag{11}$$

For the spatial error test, the expression is given by

$$\mathrm{LM}_\lambda = \frac{\left(d_\lambda - TD^{-1}d_p\right)^2}{\left[T\left(1 - TD\right)\right]} \sim \chi^2(1) \tag{12}$$

The results of the test are shown in Table 3. Based on (*26*) the robust LM test supports spatial dependence due to the omission of spatial lag.

Spatial Lag Model

Based on the previous section, the LM test supported estimating a spatial lag model. The spatial lag model implies spatial spillovers in X and the errors characterized by the expression

$$y = \left(I - \rho W\right)^{-1}\left(X\beta + \mu\right) \tag{13}$$

In Equation 13 the term $\left(I - \rho W\right)^{-1}$ is referred to as the spatial multiplier (*28*). The presence of the spatial multiplier implies global spillovers through the entire spatial system. This phenomenon was expected in this project since the number of crashes is affected not only by the immediate neighborhood but also by traffic converging from the road network to the analyzed TAZs.

For model estimation, the presence of the spatially lagged dependent variable on the right–hand side of the equation induces endogeneity or simultaneous equation bias in the spatial lag model. Therefore, OLS is inappropriate as estimator (*26*). Due to this limitation, estimation in this study is based on the application of the principles of two–stage least squares, using what is referred to as spatial two–stage least squares, or S2SLS (*29*). Standard error using White-adjusted variances was applied to handle heteroscedasticity. The results of the model are shown in Table 4.

Discussion of the Results

The results support the spatial lag model with White standard errors as the most appropriate to identify factors influencing the likelihood of pedestrian crashes in the city of Medellin, Colombia. Lagged dependent variables were tested in the model but none of them turned out to be significant. Four different categories of variables were defined in the model: sociodemographic, land use, infrastructure, and transportation mode related variables. In the original OLS all the variables were highly significant. After correcting for spatial dependence and assessing a robust estimation, some of the variables turned out to be marginally significant at 80% confidence interval.

The model indicates, in the first place, a negative effect of the variable socioeconomic status on vehicle–pedestrian collisions. As expected, zones with higher income tend to have fewer pedestrian crashes since they have a more extensive use of private motor vehicles. When analyzing the effect of the land use related variable, two opposite effects were supported by the model. In general, residential areas tend to have a reduced risk of crashes while commercial areas have an increased risk. When considering the type of road model it seems that a greater percentage of collector roads increases the likelihood of pedestrian crashes. In general, the medium traffic volumes and road width of collector roads promote mid-block crossing, leading to riskier situations for pedestrians.

Finally, three of the proxies characterizing the transportation modes were significant. These variables include the number of taxi stands, length of transit routes and parking garages per TAZ. The model supports, as does previous research, that parking garages tend to increase the number of pedestrian crashes since drivers have to walk to complete their trip. The same trend is shown by the length of transit routes. This phenomenon is exacerbated in Colombia where buses stop at the user's request without defined stops along the route. Finally, use of taxis seems to reduce the likelihood of pedestrian crashes. In general, taxis cover the entire trip

Table 4. Spatial Lag Model of Vehicle–Pedestrian Collisions in Medellin, Colombia with S2SLS Estimator, Robust White Standard Errors

Variable	Parameter estimate	Standard error	*t*-ratio	*p*-value
Constant	42.568	40.775	1.044	.2964
Sociodemographic variables				
Socioeconomic status	−41.478	14.178	−2.925	.0034
Land use				
Residential area in km^2	−96.854	56.475	−1.175	.0863
Commercial area in km^2	3848.939	2368.502	1.625	.1042
Classification of roads				
Proportion of collector roads	116.426	99.761	1.167	.2432
Transportation mode				
Length of transit routes in Km	54.834	42.602	1.287	.1981
Number of taxi stands	−54.858	27.352	−2.006	.0449
Indicator parking garages	183.979	152.622	1.205	.2280
$\rho = 0.70799$		$p = .003672$		

Table 5. Marginal Effects Spatial Lag Model Estimated via S2SLS Estimator

Variable	Direct effect	Indirect effect	Total effect
Sociodemographic variables			
Socioeconomic status	−48.214	−93.821	−142.037
Land use			
Residential area in km^2	−112.586	−219.083	−331.670
Commercial area in km^2	4,474.127	8,706.242	13,180.369
Classification of roads			
Percentage of collector roads	135.337	263.354	398.691
Transportation mode			
Length of transit routes in km	63.740	124.033	187.772
Number of taxi stands	−63.768	−124.088	−187.857
Indicator parking garages	213.863	416.158	630.021

from origin to destination, decreasing the walking exposure of the user.

The marginal effect of changes in exogenous variables can be estimated using the spatial multiplier term. Starting from the reduced Equation 13 it is possible to obtain the marginal effects expression:

$$\frac{\partial y}{\partial x_k'} = \left(I - \lambda W\right)^{-1} \beta_k \tag{14}$$

The results of the direct, indirect, and total marginal effects are shown in Table 5. The results support indirect effects or spillover effects as those having a major impact on the number of crashes. Special emphasis is given to those variables which can be affected by short– and medium–term countermeasures. Based on the results, a kilometer increase in the length of transit routes within a TAZ produces an increase of 187.7 crashes per 1,000 of population. In this case, countermeasures such as enforcing buses stopping at bus stops and improving crossings at these locations might improve these statistics. Use of taxis also seems to be an appropriate measure to reduce pedestrian crash risk. Additional taxi stands reduced the number of pedestrian crashes by 187.8 per 1,000 of population during the period of analysis, 2009 to 2016. The influence of the land use variables is also significant, specifically the effect of larger commercial areas.

Conclusions and Future Research

Collisions involving pedestrians or motorcycles are among the most dangerous in terms of number of fatalities in Colombia (*1*). By providing a better understanding of the factors influencing pedestrian crashes, specific measures can be implemented to reduce the risk. Based on the analysis, the spatial dependence of the model could be determined by mapping crash locations to spatial units of analysis.

Previous research has applied both non-spatial and spatial techniques to examine the factors influencing pedestrian crashes in urban areas. Limited research has been conducted in developing countries where lack of data restricts these types of analysis. To the best of the authors' knowledge, this is the first model provided for analyzing factors influencing vehicle–pedestrian collisions in Colombia.

There are various specification search strategies in spatial econometric modeling. The reviewed literature supported a better performance with the classic approach. One of the most important components in cross-sectional analysis is to define spatial weights. Queen contiguity was used to build the matrix characterizing the spatial morphology of the analysis. Spatial diagnostics in this specification strategy supported estimation of a spatial lag model. S2SLS method with White standard errors was used for the modeling estimation.

The preliminary results of the model indicate a heterogeneous effect of land use, in accordance with the literature. In general, TAZs with bigger commercial areas have a higher rate of vehicle–pedestrian collisions. On the other hand, residential zones along with higher socioeconomic status produce a negative effect on pedestrian crashes. The model also confirmed the necessity of enforcing bus stops which can reduce the effect of length of transit routes on the rate of pedestrian crashes. Additional countermeasures include taxi parking to cover last mile transit and to enhance pedestrian infrastructure close to parking garages.

Future research should include other variables in the analysis, including exposure regarding pedestrian counts and volumes. In addition, pedestrian policies and other regulations should be revisited to determine if they are effective in reducing the pedestrian crash rates. The study developed a framework that may be further extended to other fields. It also may prove useful to agencies and groups working to reduce vehicle–pedestrian collisions by allowing them to target their efforts better. The results also demonstrate the usefulness of spatial techniques in the traffic safety field.

Acknowledgments

The authors acknowledge the support of the Center for Road Safety at Purdue University, with special thanks to Professor Andrew Tarko. We extend our appreciation to Dr Mario Romero for his invaluable contribution in the geocoding process. Finally, we wish to express our gratitude to the Secretaria de Movilidad de Medellin.

References

1. Vargas, D. Comportamiento de muertes y lesiones por accidentes de transporte. Grupo Centro de Referencia Nacional sobre Violencia Instituto Nacional de Medicina Legal y Ciencias, 2015, pp. 481-546.
2. Lord, D., and F. Mannering. The Statistical Analysis of Crash-Frequency Data: A Review and Assessment of Methodological Alternatives. *Transportation Research Part A: Policy and Practice*, Vol. 44, No. 5, 2010, pp. 291–305.
3. Loukaitou-Sideris, A., R. Liggett, and H. G. Sung. Death on the Crosswalk. *Journal of Planning Education and Research*, Vol. 26, No. 3, 2007, pp. 338–351. http://journals.sagepub.com/doi/10.1177/0739456X06297008.
4. Siddiqui, C., M. Abdel-Aty, and K. Choi. Macroscopic Spatial Analysis of Pedestrian and Bicycle Crashes. *Accident Analysis & Prevention*, Vol. 45, 2012, pp. 382–391.
5. Wier, M., J. Weintraub, E. H. Humphreys, E. Seto, and R. Bhatia. An Area-Level Model of Vehicle-Pedestrian Injury Collisions with Implications for Land Use and Transportation Planning. *Accident Analysis & Prevention*, Vol. 41, No. 1, 2009, pp. 137–145.
6. Lee, C., and M. Abdel-Aty. Comprehensive Analysis of Vehicle-Pedestrian Crashes at Intersections in Florida. *Accident Analysis & Prevention*, Vol. 37, No. 4, 2005, pp. 775–786.
7. Brugge, D., Z. Lai, C. Hill, and W. Rand. Traffic Injury Data, Policy, and Public Health: Lessons from Boston Chinatown. *Journal of Urban Health*, Vol. 79, No. 1, 2002, pp. 87–103.
8. LaScala, E. A., D. Gerber, and P. J. Gruenewald. Demographic and Environmental Correlates of Pedestrian Injury Collisions: A Spatial Analysis. *Accident Analysis & Prevention*, Vol. 32, No. 5, 2000, pp. 651–658.
9. Miranda-Moreno, L. F., P. Morency, and A. M. El-Geneidy. The Link Between Built Environment, Pedestrian Activity and Pedestrian-Vehicle Collision Occurrence at Signalized Intersections. *Accident Analysis & Prevention*, Vol. 43, No. 5, 2011, pp. 1624–1634. http://dx.doi.org/10.1016/j.aap.2011.02.005.
10. Schneider, R., M. Diogenes, L. Arnold, V. Attaset, J. Griswold, and D. Ragland. Association Between Roadway Intersection Characteristics and Pedestrian Crash Risk in Alameda County, California. *Transportation Research Record: Journal of the Transportation Research Board*, 2010. 2198: 41–51.
11. Ewing, R., and E. Dumbaugh. The Built Environment and Traffic Safety: A Review of Empirical Evidence. *Journal of Planning Literature*, Vol. 23, No. 4, 2009, pp. 347–367.
12. Kim, K., I. M. Brunner, and E. Yamashita. Modeling Fault Among Accident-Involved Pedestrians and Motorists in Hawaii. *Accident Analysis & Prevention*, Vol. 40, No. 6, 2008, pp. 2043–2049.
13. Ukkusuri, S., L. F. Miranda-Moreno, G. Ramadurai, and J. Isa-Tavarez. The Role of Built Environment on Pedestrian Crash Frequency. *Safety Science*, Vol. 50, No. 4, 2012, pp. 1141–1151.
14. Ukkusuri, S., S. Hasan, and H. Aziz. Random Parameter Model Used to Explain Effects of Built-Environment Characteristics on Pedestrian Crash Frequency. *Transportation Research Record: Journal of the Transportation Research Board*, 2011. 2237: 98–106.
15Narayanamoorthy, S., R. Paleti, and C. R. Bhat. On Accommodating Spatial Dependence in Bicycle and Pedestrian Injury Counts by Severity Level. *Transportation Research Part B: Methodological*, Vol. 55, 2013, pp. 245–264.
16. Cottrill, C. D., and P. Thakuriah. Evaluating Pedestrian Crashes in Areas with High Low-Income or Minority Populations. *Accident Analysis & Prevention*, Vol. 42, No. 6, 2010, pp. 1718–1728.
17. Jermprapai, K., and S. Inivasan. Planning-Level Model for Assessing Pedestrian Safety. *Transportation Research Record: Journal of the Transportation Research Board*, 2014. 2464: 109–117.
18. Chakravarthy, B., C. L. Anderson, J. Ludlow, S. Lotfipour, and F. E. Vaca. The Relationship of Pedestrian Injuries to Socioeconomic Characteristics in a Large Southern California County. *Traffic Injury Prevention*, Vol. 11, No. 5, 2010, pp. 508–513.
19. McArthur, A., P. T. Savolainen, and T. J. Gates. Spatial Analysis of Child Pedestrian and Bicycle Crashes Development of Safety Performance Function for Areas Adjacent to Schools. *Transportation Research Record: Journal of the Transportation Research Board*, 2014. 2465: 57–63.

20. Green, J., H. Muir, and M. Maher. Child Pedestrian Casualties and Deprivation. *Accident Analysis & Prevention*, Vol. 43, No. 3, 2011, pp. 714–723.
21. LeSage, J., and R. K. Pace. Spatial Econometric Models. *Handbook of Applied Spatial Analysis*, No. 2007, 2010, pp. 355–376.
22. Steer Davis Gleave. *Estudios de movilidad para los macroproyectos de las centralidades norte y sur del Valle de Aburra*, 2011.
23. Qin, X., J. N. Ivan, and N. Ravishanker. Selecting Exposure Measures in Crash Rate Prediction for Two-Lane Highway Segments. *Accident Analysis & Prevention*, Vol. 36, No. 2, 2004, pp. 183–191.
24. Florax, R. J. G. M., H. Folmer, and S. J. Rey. A Comment on Specification Searches in Spatial Econometrics: The Relevance of Hendry's Methodology: A Reply. *Regional Science and Urban Economics*, Vol. 36, No. 2, 2006, pp. 300–308.
25. Hendry, D. F. Econometrics: Alchemy or Science. *Economica*, Vol. 47, No. 188, 1980, pp. 387–406.
26. Anselin, L., and S. Rey. *Modern Spatial Econometrics in Practice: A Guide to GeoDa, GeoDaSpace and PySAL*. GeoDa Press LLC, Illustrate, 2014.
27. Anselin, L., A. K. Bera, R. Florax, and M. J. Yoon. Simple Diagnostic Tests for Spatial Dependence. *Regional Science and Urban Economics*, Vol. 26, No. 1, 1996, pp. 77–104.
28. Anselin, L. Spatial Externalities, Spatial Multipliers, and Spatial Econometrics. *International Regional Science Review*, Vol. 26, No. 2, 2003, pp. 153–166.
29. Kelejian, H. H., and I. R. Prucha. A Generalized Spatial Two-Stage Least Squares Procedure for Estimating a Spatial Autoregressive Model with Autoregressive Disturbances. *Journal of Real Estate Finance and Economics*, Vol. 171, 1998, pp. 99–121.

The Standing Committee on Transportation in the Developing Countries (ABE90) peer-reviewed this paper (18-01753).

Article

TRR
JOURNAL OF THE TRANSPORTATION RESEARCH BOARD

Transportation Research Record
2018, Vol. 2672(31) 63–75

DOI: 10.1177/0361198118758652
journals.sagepub.com/home/trr
SAGE

Using Open Source Data to Measure Street Walkability and Bikeability in China: A Case of Four Cities

Peiqin Gu[1], Zhiyuan Han[1], Zhejing Cao[2], Yulin Chen[2], and Yang Jiang[1]

Abstract
Whether people would choose to walk or ride a bike for their daily travel is affected by how desirable the environment is for walking and biking. To better inform urban planning and design practices, studies on measuring walkability and bikeability have emerged in western countries over the last decade. However, such efforts are still rare in developing countries, partially due to the scarcity of urban data. Utilizing open source data, this paper puts forward a methodology to comprehensively and objectively measure street walkability and bikeability in China. The methodology was applied to four Chinese cities: Tianjin, Chongqing, Kunming, and Shijiazhuang. Analyses showed the following results: (1) city centers tend to have higher walkability than periphery areas; (2) a preliminary bikeable street network exists in most cities (except mountainous cities), but the prevalence of bike lanes on streets is much lower than that of sidewalks; (3) the problem of illegal parking on both sidewalks and bike lanes is severe, especially in city centers; (4) biking safety and comfort is compromised due to a lack of physically separated bike lanes; and (5) the street wall continuity varies from place to place whereas the street network in traditional city centers is much denser than newly developed car-oriented areas. The end of the paper provides corresponding policy implications.

As car-oriented urban development during past decades has caused severe traffic jams and air pollution in China, the public is starting to realize that streets are an important public space in daily life and is paying more attention to the non-motorized transportation environment. With the emerging popularity of bike-sharing businesses in China, the mismatch between biking demand and the supply of street space for biking has never before been so striking. In fact, the central government of China published *China's Urban Pedestrian and Bicycle Transportation System Planning and Design Technical Guideline* in 2013 (*1*). Chinese cities have also turned their attention to non-motorized transportation very recently; for instance, Shanghai published the *Shanghai Street Design Guideline* in 2016 (*2*).

In theory, whether people choose to walk or ride a bike for their daily travel should be affected by how desirable the environment is for walking and biking. Although the literature has supported this with rich empirical evidence, it is often limited in the following aspects: the coding of comprehensive street design could only be done at a small scale due to labor intensity (*3, 4*); the crowd-sourcing of residents' opinions on the walk and bike environments are criticized as lacking objectivity (*5, 6*); the existing street score evaluations tend to ignore detailed design components (*7, 8*); there are still few studies measuring street bikeability (*9, 10*); and there are still limited approaches that can be linked with the Chinese urban context (*11–13*). In the practical world, urban planners and policy makers are urging a comprehensive and objective approach to measuring walkability and bikeability that can guide real-world planning and design exercises.

How can we massively quantify street planning and design indicators to measure street environments for walking and biking in China? To answer this question, a methodology utilizing open source data was put forward to comprehensively and objectively measure street walkability and bikeability in China. The methodology was applied to four Chinese cities: Tianjin, Chongqing, Kunming, and Shijiazhuang. The remainder of the paper is organized as follows. First existing street scoring methodologies and their applications are reviewed. The paper then explains how all individual indicators were processed and how the framework was developed using the information entropy-weighting method. Next, the overall score results and spatial patterns of individual indicators for the four cities are illustrated. The paper concludes with the findings and policy recommendations.

[1]China Sustainable Transportation Center, Beijing, China
[2]School of Architecture, Tsinghua University, Beijing, China

Corresponding Authors:
Address correspondence to Yulin Chen: chenyulin@mail.tsinghua.edu.cn;
Yang Jiang: yangjiang@chinastc.org

Literature Review

The street environment impacts non-motorized transportation trip generation and human physical activities. A quantifiable environmental evaluation framework and index for non-motorized travel is crucial for precise design and detailed management for urban planners, designers, and regulators. In recent years, the Semantic Differential method by Osgood et al. (*3*) and the Public Space and Public Life method by Jan Gehl (*4*) have been widely adopted for the assessment of public spaces and walking environments. However, these methods fall short of massive-scale evaluation due to their high labor cost. Alternative crowd-sourcing approaches, such as "Rate my Street" (*5*) and NEW-Y survey (*6*), have prevailed. Nevertheless, these methods often lack reliability due to the non-uniform standards and the missing of feedback data in blind-spot areas. More importantly, perceived assessment from crowd-sourcing overlooks the objective dimension of urban form. In fact, studies have shown that objective measures of the built environment had stronger associations with walking (*14*), and they were widely applied to test the walkability of neighborhoods influencing human behavior and perception (*15–18*).

Walkability assessment tools have emerged in urban planning and transport-related research, as well as health-related research. For example, Frank et al. (*7, 8*) proposed a walkability index based on indicators such as land-use mix, residential density and street connectivity, and explored its link with human physical activities, obesity, air quality and neighborhood quality of life. However, they focused mostly on facility accessibility without considering street design quality or the human-scale environment. Dyck et al. (*19*) defined neighborhood walkability, both by geographical map data and observations. Millington et al. (*20*) developed the Scottish Walkability Assessment Tool to record physical environment aspects of functionality, safety, aesthetics, and destination; this was more comprehensive than the previous walkability index. The Walk Score® company put forward an approach to calculate facility accessibility on a large scale from point-of-interest data with distance decay and neighborhood dimensional effects; however, they also failed to quantify more human-scale variables (*21*).

In recent years, some Chinese scholars started to further develop walkability assessment in local contexts. Lu (*11*) established a walkability evaluation framework based on use frequency, use variety and distance to community facilities by walking. Wu et al. (*12*) found that micro-blog check-in frequency correlates to walk score based on 15 kinds of life service facilities in Shenzhen. Some other scholars incorporated physical spatial features into the walk score. Liu et al. (*13*) evaluated the comforts of the walking environment of a historical district in Tianjin by creating a 3-dimension index of 15 factors, including street space comforts, physical environment comforts, and street furniture comforts. However, there is still a great need for combining features of both facility accessibility and street environments, especially in the context of Chinese cities. For example, many sidewalks are occupied by parked vehicles, and people are exposed to high danger when crossing the streets in suburban areas due to a lack of crosswalks.

In addition, the cycling environment should also be considered in cities with high levels of cycling. However, research on cycling is less extensive and appears to focus mostly on the compatibility of roadways for bicycles and motor vehicles. Some exceptions are that Pikora et al. (*9*) developed a framework for the assessment of environmental determinants of walking and cycling, and Winters et al. (*10*) proposed the bike score based on mapping bike lanes, slopes, origin-destination pairs and shared parking facilities. Research in these areas is still insufficient.

To summarize, studies and tools on measuring walkability and bikeability have emerged in western countries over the last decade to better inform urban planning and design practices. However, such efforts are still rare in developing countries, partially due to the scarcity of urban data. A more comprehensive street score framework is needed to bridge the existing research gaps. Thus, in this paper, we propose a street-score framework that can do the following:

(1) overcome subjectivity by observing the built environment objectively;
(2) quantify not only street accessibility but also design factors for safety and comfort;
(3) be scalable to cover most built areas when possible;
(4) weight cycling environments together with pedestrian environments;
(5) include indicators that correspond to local context;
(6) be simple enough to further provide implications for urban and transportation planners, designers, and regulators.

Data and Methodology

Data Collection

In this paper, four Chinese cities, Tianjin, Chongqing, Shijiazhuang, and Kunming, were used in our street score experimentation. As shown in Table 1, these four cities cover different economic and urban development stages, geographic locations, terrain factors, and degrees of public transportation infrastructure. This paper does not include other smaller cities due to low availability of street images.

To make our method adaptable for other cities, we depended on open source web API providers. Basic street network vector data were extracted from openstreetmap.org on 2017. We collapsed dual lines representing 2-way lanes into street center lines and corrected street topology to gain a total of 19,767 street segments and 6,702 km: 1,590 km for Tianjin, 1,395 km for Chongqing, 2,393 km for Kunming, and 1,324 km for Shijiazhuang. These streets covered central

Table 1. Key Characteristics of the Four Cities in the Study

	Population (million)	Built-up area (km^2)	GDP (yuan)	Location	Terrain type	Metro lines in operation
Tianjin	9.4	885.4	143129	Northern	Plain	4
Chongqing	19.5	1329.5	37691	Southern	Mountainous	5
Kunming	4.1	420.5	62437	Southern	Plateau	2
Shijiazhuang	2.6	278.1	48491	Northern	Plain	0

Note: Data from Chinese Urban-Rural Construction Statistical Yearbook 2015.

urban areas; other streets were not included due to low data availability. Then, we divided each street segment by 50 m and generated a total of 37,380 interval points (those with a length less than 50 m generated one point), which were spatially linked to their original segments.

High-resolution street view imagery has been recently used as an important source for street environment evaluation (*22–25*). In this paper, we used the Tencent street view imagery service (map.qq.com) to collect street images for these four cities. Images taken during both the spring and summer periods for the same street were collected to avoid potential bias caused by seasonal factors. Street images used for Tianjin were originally taken during August 2014, Shijiazhuang during June 2015, Kunming during June 2014, and Chongqing during April 2015. Utilizing the API, images were obtained in four different directions at positions of 0, 90, 180, and 270 degrees for each interval point, accumulating a total of 131,048 images: 30,492 for Chongqing, 46,252 for Kunming, 20,112 for Shijiazhuang, and 34,192 for Tianjin. These images were linked back to street segments via the established spatial relationship between interval points and street segments. Via these images, we extracted information such as right of way, tree shade, and car parking for each street segment. Moreover, we also geocoded street crossing facilities in three categories: at-grade striped crossings, underpass crossings, and overpass crossings.

Point of interest (POI) data from Gaode (amap.com) was also collected, yielding a total of 1,194,125 points. These included categories such as residential neighborhoods, governmental services, research and education institutions, medical and health institutions, retail stores, entertainment centers, companies, transportation facilities, logistics warehouses, green spaces and parks. Building footprint raster data was extracted from Baidu (map.baidu.com), and then vectorized to get a total of 456,402 building lots: 173,781 for Chongqing, 63,152 for Kunming, 80,878 for Shijiazhuang, and 138,591 for Tianjin. All these data were adjusted into the standard WGS84 coordinate system, and were geo-processed under corresponding Gaussian projection systems.

Street Score Framework for Walkability and Bikeability

By referring to the three main principles put forward in *China's Urban Pedestrian and Bicycle Transportation System Planning and Design Technical Guideline* (*1*), safety, continuity, and convenience, and Jen Gehl's criteria for public space quality in protection, comfort, and enjoyment (*4*), we extended the existing framework to evaluate street walkability and bikeability from three aspects, namely safety, comfort, and convenience. This framework was designed to be closely connected to China's urban transportation planning and design context, and we hoped to provide guidance for real world practices.

Regarding safety, we considered sidewalks, bike lanes, and crossing facilities as the basic security for pedestrians and bikers. Meanwhile, illegal car parking on sidewalks or bike lanes will lead to potential security problems; thus, we calculated sidewalk existence, crossing facility existence, and the ratio of sidewalks to illegal parking, to measure a street's "safety" score for walking. We also calculated bike lane existence, crossing facility existence, and the ratio of bike lanes to illegal parking, to measure a street's "safety" score for biking.

As for comfort, contrary to the car-oriented wide roads that have been paved in new towns over the past decades, we conceived that narrower streets with continuous façades could provide a sense of human scale (*23*). Streets with enough tree shade are more appealing to pedestrians who are not willing to get sunburn. Therefore, we calculated street width, the street–wall continuity ratio, and the percentage of streets with tree shade, to represent the comfort of pedestrians. As for biking, a bike lane with proper separation from automobiles secures the continuity of biking right-of-way, and tree shading lowers temperatures. Thus, we mainly considered bike lane isolation and the percentage of streets with tree shade to assess "comfort".

When we speak of convenience for non-motorized transportation, it should reflect how streets can better connect pedestrians and bikers to their daily destinations. Neighborhoods with denser street networks, crossings, and abundant facilities would shorten residents' travel distance. Thus, in this paper, we calculated street network density, street crossing density, and a facility accessibility index to measure street "convenience" for walking and biking. The scoring framework for street walkability and bikeability is shown in Table 2.

Table 2. Street Score Framework

<table>
<tr><th rowspan="2">Dimension</th><th colspan="2">Indicator[a]</th></tr>
<tr><th>Walkability</th><th>Bikeability</th></tr>
<tr><td>Safety</td><td>Sidewalk existence X_1 (+)</td><td>Bike lane existence X_2 (+)</td></tr>
<tr><td></td><td colspan="2">Crossing facility existence X_3 (+)</td></tr>
<tr><td></td><td>% Sidewalk with illegal parking X_4 (−)</td><td>% Bike lane with illegal parking X_5 (−)</td></tr>
<tr><td>Comfort</td><td colspan="2">% Street with tree shade X_6 (+)</td></tr>
<tr><td></td><td>Street width X_7 (−)</td><td>Bike lane isolation X_8 (+)</td></tr>
<tr><td></td><td>Street wall continuity ratio X_9 (+)</td><td></td></tr>
<tr><td>Convenience</td><td colspan="2">Street network density X_{10} (+)</td></tr>
<tr><td></td><td colspan="2">Crossing facility density X_{11} (+)</td></tr>
<tr><td></td><td colspan="2">Facility accessibility X_{12} (+)</td></tr>
</table>

[a]Indicators marked as '+' imply a positive relationship with street score, while indicators marked as '−' imply a negative relationship with street score.

We produced dummy variables for each street segment, reflecting the existence of sidewalks, bike lanes, and crossing facilities, where 1 represents existence whereas 0 represents none. Through street images, we estimated the proportions of sidewalks and bike lanes (in length) occupied by illegal car parking, as well as the proportion of streets (in length) with tree shading. We determined the street facade, calculated the street wall continuity ratio, and determined the street width as the distance from the street's center line to the building facades on both sides. In general, there are three forms of bike lane isolation in China: stripes, fences, and medians. This category is coded in an ordered score fashion to represent the level of comfort it can provide for bikers: streets with no bike lane at all are scored with 0; stripes are not physical isolation, but they can at least provide bikers with a right-of-way, and thus streets with stripes are scored with 1; fences provide effective physical isolation, and thus streets with fences are scored with 2; and medians are scored with 3 because they bring more shading and physical isolation to protect bikers.

Based on interval points, street network density was calculated as the length of a street by a 1 km^2 buffer zone of interval points; crossing density was calculated as the ratio of the number of crossing facilities to the length of streets within the 1 km^2 buffer zone. As for the facility accessibility index, which was adapted from the Walk Score calculation in several Chinese cities (*11–13*), we identified and extracted seven daily facilities (farmers' market, convenience store, restaurant, supermarket, leisure, bank, and hospital) and two public transportation facilities (regular bus stop, and metro station) from the POI database. Then, using network analysis tools, we calculated the nearest street network distance from a certain interval point to each of these facilities as $\mathrm{Dis_p}$, where p includes nine POI categories. Adapting the distance decay function developed by Wang et al. (*26*), the accessibility index to each of these facilities $\mathrm{Acc_p}$ was calculated as follows:

$$\mathrm{Acc_p} = \begin{cases} 0, \text{if } \mathrm{Dis_p} > 2400 \\ 1, \text{if } \mathrm{Dis_p} \le 400 \\ 188177 * \dfrac{\mathrm{Dis_p}}{(\mathrm{Pow}(\mathrm{Dis_p}, 2.9475) + 34259163)}, \text{if } 400 < \mathrm{Dis_p} < 2400 \end{cases} \tag{1}$$

Then, the average facility accessibility for each interval point was calculated as follows:

$$\mathrm{Acc} = \sum_{1}^{9} \mathrm{Acc_p} * 1/9 \tag{2}$$

Finally, attaching these three point-based indicators to a street segment, we calculated the average street network density, crossing density and facility accessibility index.

Using Information Entropy Weighting to Derive Scores

Shannon's Information Entropy Weighting (*27*) method was adopted in this paper to derive the final street score for walkability and bikeability, as well as scores in sub-dimensions. This method was chosen because it can overcome the subjectivity of weights from artificial and arbitrary effects, and can avoid the overlapping of information among multiple variables (*28–32*). The specific calculation steps are as follows.

First, the data matrix $X = \{X_{ij}\} \mathrm{m} \times \mathrm{n} (0 \le \mathrm{i} \le \mathrm{m}, 0 \le \mathrm{j} \le \mathrm{n})$ composing of m streets and n indicators was built, where X_{ij} is the value of the j th indicator of the i th street.

Second, since different indicators differentiate in terms of dimension and unit, we computed the dimensionless X'_{ij} for each indicator using the following equations:

$$X'_{ij} = \left(X_{ij} - \min\{X_j\}\right) / \left(\max\{X_j\} - \min\{X_j\}\right) \quad (3)$$

$$X'_{ij} = \left(\max\{X_j\} - X_{ij}\right) / \left(\max\{X_j\} - \min\{X_j\}\right) \quad (4)$$

Here, the indicators meeting the requirements of the positive and negative hypotheses in Table 2 were calculated with Equation (3), while the indexes meeting the requirements of the negative hypothesis were calculated with Equation (4).

In the third step, we calculated the proportion of the index R'_{ij} :

$$R'_{ij} = X'_{ij} / \sum_{i=1}^{m} X'_{ij} \quad (5)$$

The fourth step was to calculate the entropy of the j th index e_j :

$$e_j = \frac{1}{\ln(n)} * \sum_{i=1}^{m} R'_{ij} \ln\left(R'_{ij}\right) \quad (6)$$

If $R'_{ij} = 0$, then $\ln\left(R'_{ij}\right)$ was deemed as 0.

In the fifth step, we calculated the imbalance coefficient g_j of the j th indicator:

$$g_j = 1 - e_j \quad (7)$$

The sixth step was to calculate the weight of the j th indicator W_j :

$$W_j = \left(g_j\right) / \sum_{j}^{n} (\left(1 - g_j\right) \quad (8)$$

In the seventh step, we calculated the value of the j th indicator of the i th city V_{ij} of:

$$V_{ij} = \sum_{j}^{n} W_j X'_{ij} \times 100 \quad (9)$$

Then, the evaluation indicator V_i of the i th street was calculated as follows:

$$V_i = \sum_{j}^{n} V_j \quad (10)$$

This value was further standardized into a 0–100 scale:

$$V_i = \left(V_i - \min\{V_i\}\right) / \left(\max\{V_i\} - \min\{V_i\}\right) \quad (11)$$

Finally, the street scores for walkability V^{w}_{i} and bikeability V^{B}_{i} for the i th street were calculated as follows:

$$V^{w}_{i} = \sum_{j}^{p} V_j, p \in \{X_1, X_3, X_4, X_6, X_7, X_9, X_{10}, X_{11}, X_{12}\} \quad (12)$$

$$V^{B}_{i} = \sum_{j}^{q} V_j, q \in \{X_2, X_3, X_5, X_6, X_8, X_{10}, X_{11}, X_{12}\} \quad (13)$$

Table 3. Street Score Results of the Four Cities

	City				
	Tianjin	Chongqing	Kunming	Shijiazhuang	Total
Walk Score	59	54	48	47	52
Safety	79	74	63	64	70
Convenience	49	46	46	37	46
Comfort	52	48	45	53	49
Bike Score	47	35	38	37	40
Safety	70	56	58	61	61
Convenience	48	45	45	36	45
Comfort	30	15	22	29	23

Sets of indicators with subscripts are listed in Table 2.

Results

Walkability and Bikeability in Four Chinese Cities

Street Score Overview. Based on the proposed methodology, the street score framework was applied to Tianjin, Chongqing, Kunming, and Shijiazhuang. Table 3 summarizes the average walk scores, bike scores, and sub-scores. Figure 1 shows the walk and bike score of every street for the four cities. The following similarities were found between the four cities, which correspond with intuitive understandings. First, in all cities, scores on safety are the highest among three aspects, suggesting that city governments are prioritizing basic infrastructure for walking and biking. Second, as seen in Figure 1*a*, all cities formed clusters of high-walkability streets, contributing to continuous pedestrian systems where citizens are more likely to occupy; meanwhile, in peripheral areas, street walk scores faded out. Third, Figure 1*b* shows a pattern of high bikeability street networks in Tianjin, Kunming, and Shijiazhuang despite the overall score being low compared to the walk score.

Looking at single cities in detail, Tianjin stands out in both walk and bike scores because its residents benefit from a legacy of an enclave urban structure, where diversity is high enough to encourage daily travel by foot and bike. According to statistics from the Tianjin Urban Planning and Design Institute, the share of cycling among all transport modes in Tianjin is 33% (*33*), which is significantly higher than in large international cities such as Beijing, Shanghai, London, and Seoul (*34–37*). Conversely, Chongqing lacks biking facilities due to its steep terrain; however, its walk score is high because the mountain has divided the city into clusters with diverse urban functions, which suits walking.

Evaluation of Safety/Comfort/Convenience. Sub-scores also presented different patterns across the cities since geographic location and development stages varied. The frequency distribution of each sub-score was examined by breaking the scores into 10 intervals to determine the safety, comfort, and convenience (see Figure 2).

红桥区
河北区
南开区
河东区
Tianjin
东丽区
JINGJINTANG EXPY
JINBIN AVE
李楼
双港镇
江北区
YUYI EXPY
Chongqing
沙坪坝区
九龙坡区
大渡口区
Dadukou Qu
Taoyuan
Doupo
KUNQU EXPY
Longtou
Zhuangke
Panlong
Kunming
Ala
Xiaobanqiao
Guandu
Shangjing Village
Xizhaotong Town
Shijiazhuang
长安区
桥西区
裕华区
Beidou Village
East Yangling Village
Bafang Village
Liu Village
Fangcun Town
Xima Town
XiLiangzheng Village
Dongyin Village
Dongzuo Village
0 20 40 60 80 100

(a)

(continued)

(b)

Figure 1. Street score map over four cities in which (*a*) shows walk score, and (*b*) shows bike score.

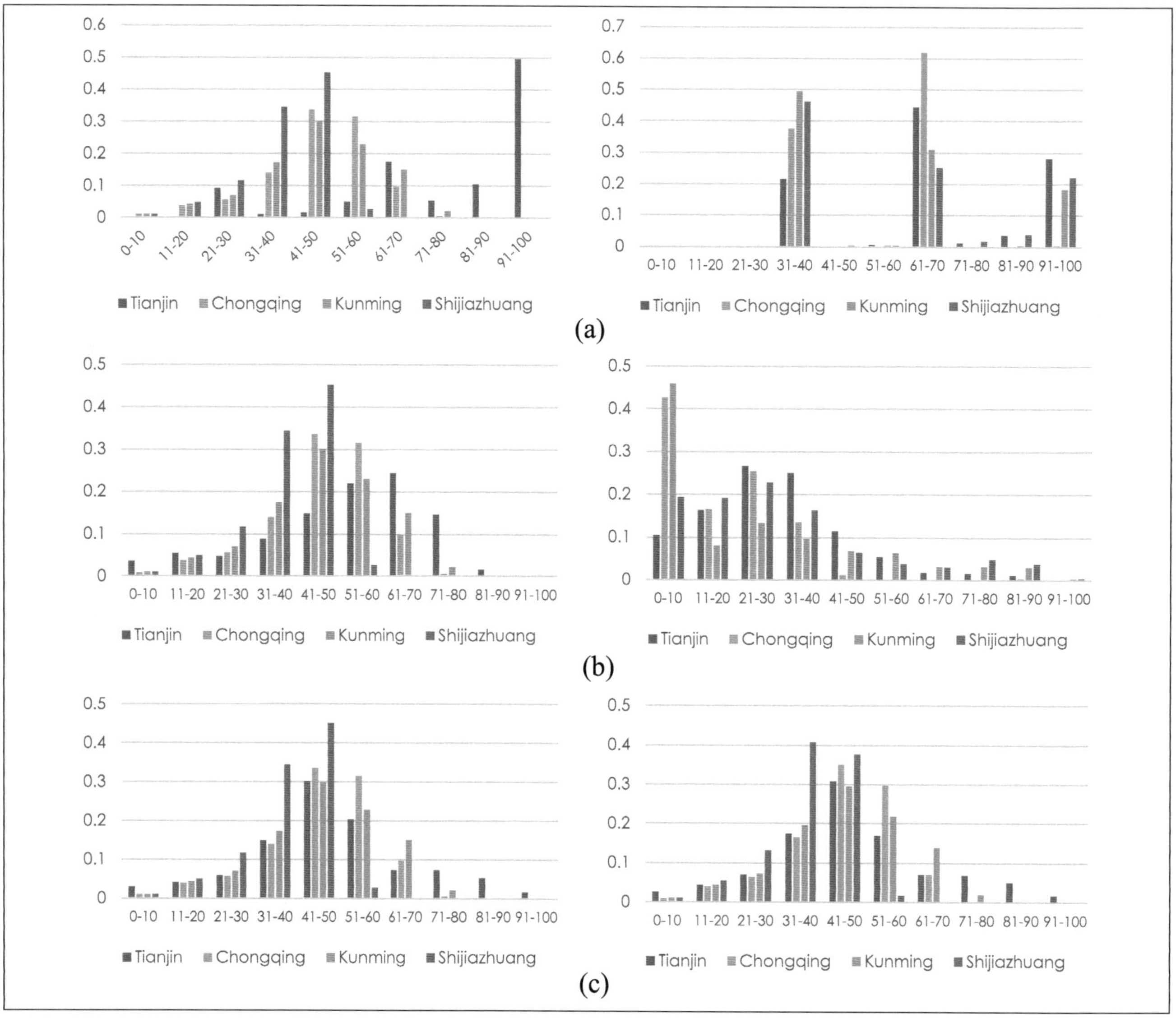

Figure 2. Frequency distribution of street walk sub-scores (left) and bike sub-scores (right) for (*a*) safety, (*b*) comfort, and (*c*) convenience.

Walking safety in Tianjin stood out, with the peak value over 90, while the same score for other cities remained aggregated around the middle, indicating that quite a few streets in Tianjin provide a safer pedestrian environment. For cycling safety, most scores were concentrated in the ranges of 31–40, 61–70, and above 90. Since the indicators used to measure cycling safety were mostly binary variables, the scores had similar values. Chongqing showed the expected low scores for bikeability due to steep terrain, but the other three cities had cycling safety scores close to each other.

Walking comfort presented a pattern similar to convenience across the four cities, except that Tianjin did not earn as many high scores. Cycling comfort exhibited an interesting pattern, in that the scores were concentrated around 30, with a lowest average among all sub-scores of 23. A possible reason for this is a lack of dedicated bike lanes in most areas across the cities. Therefore, it is urgent for cities to bring more bike lanes separated from motorized traffic, and to encourage more cycling by promoting comfort and reinforcing safety.

Convenience is mostly determined by the pattern of urban form and the supply of daily service facilities. The score of convenience across the four cities was more normally distributed than the other scores, which were concentrated around their mean. For both walking and cycling, a majority of streets in Shijiazhuang earned a score under 50, while streets with high scores over 70 mostly came from Tianjin.

Comparative Analysis of Indicators

Indicators Overview. While the street scores and sub-scores measure the comprehensive condition of the pedestrian and cycling environment, the individual indicator generates practical benchmarks for real world urban and transportation planning and design exercises. The proposed street score framework consists of a series of indicators that examine specific street-related issues, covering the structure of street network, street facilities, the spatial relationship between streets and buildings, as well as illegal parking. Improving each of these indicators would contribute to a more convenient, safe, and comfortable pedestrian and cycling system.

Sidewalk Existence. The existence of sidewalks is definitely a key contributor to a safe walking environment. Figure 3*a* shows sidewalk distribution over four cities. Overall, 85% of the streets in length in Tianjin are built with a sidewalk either on one side or on both sides. With such high coverage of sidewalks, the pedestrian system in Tianjin is perfectly connected. With a coverage of 68%, Chongqing comes in second place, where the north part of the city looks as good as Tianjin regarding the distribution of sidewalks; this area happens to be the most vibrant area of Chongqing. Shijiazhuang had lower sidewalk coverage, especially in periphery industrial zones. Urban planners should pay attention to better industrial and urban integration; adding more sidewalks around industrial workplaces might be a necessary first step.

Bike Lane Existence. In terms of the cycling environment, the existence of bike lanes is a key indicator of cycling safety. As quantities of shared bikes are gradually bringing bicycles back to the city after motorized traffic took control of urban roads, designated bike lanes have turned out to be of tremendous necessity. In Tianjin and Shijiazhuang, over one-third of the streets have bike lanes; there are no bike lanes across the city of Chongqing due to the mountainous terrain. Although the coverage of bike lanes is close in Tianjin and Shijiazhuang, the spatial pattern is significantly different. As shown in Figure 3*b*, a highly connected and continuous cycling network has formed in the northern part of central Tianjin; the pattern is more or less random in Shijiazhuang. A bike lane network with good connectivity and accessibility is no doubt the first step to achieving a green transportation development goal. In Kunming, even though the coverage is relatively lower, bike lanes are mainly constructed along primary and secondary roads; this is a smart move toward a thoroughly distributed cycling network.

Sidewalks with Illegal Parking. With increasing motorized traffic, parking lots have become a limited resource at major urban destinations and around residential zones, which leads to many illegally parked vehicles occupying sidewalks, bike lanes and even driveways. Therefore, illegal parking on sidewalks is another important measurement of walking safety. As shown in Figure 3*c*, Tianjin suffers the most from illegal parking on sidewalks; a large area of central Tianjin is confronted with the challenge to optimize parking space and remove illegal parking so as to protect pedestrians from potential danger. Similarly, almost the whole center of Shijiazhuang presents a high proportion of illegal parking on sidewalks. In contrast, Kunming exhibits little worry about such issue, even in the crowded old town. Chongqing has just begun to present illegal parking issues in a few commercial centers and residential zones nearby; this should act as a warning to the government and planning institutes, and prompt timely regulation. To address illegal parking in motorizing cities, better enforcement of parking regulation and smart monitoring are a necessity without doubt. Such an issue could be further ameliorated by introducing policies like congestion charging to reduce car use in core areas with limited parking spaces.

Street Wall Continuity Ratio. The street wall continuity ratio measures the continuity of building façades along streets, which is an important determinant of a comfortable walking experience. Tianjin exhibited an average street-wall continuity ratio of 36%, more than double that of Shijiazhuang, and much higher than that of Chongqing and Kunming. As shown in Figure 3*d*, high-value streets cover a large area in central Tianjin, contributing to a well-organized building facade along streets. Meanwhile, these high-value streets are distributed in clusters, providing persistent comfort to the walking experience. Kunming actually presents a high street-wall continuity ratio in the old town area. Nonetheless, the high-value streets are not systematically connected in most cases.

Street Network Density. Street network density, which is closely related to block size, is a basic but important indicator of walkability and bikeability; it has been repeatedly verified that a dense street network and small blocks considerably improve citizens' walking and cycling experience (*38, 39*). It turns out to be more convenient to reach a certain destination if the network is designed in a walkable scale. Due to different development histories and geographic terrains, the four cities formed distinct street network structures. The spatial pattern of the street network density is visualized in Figure 3*e*. It is obvious that the old town of Kunming presents the highest density, with most areas exceeding 8 km/km^2, a target set by the Chinese central government (*40*). Tianjin comes in next, with an average density of 6.7 km/km^2. A well-organized street network with high density covers a large area in central Tianjin along the main river running through the city, which turns out to be the most walkable and attractive site in Tianjin. This is where the concession area is located, a historical site constructed during the 1860s–1940s. The urban structure has been preserved since then, and it is even taken as an example for the preservation of historical and cultural towns because of the fine scale of blocks as well as the well-designed and ordered buildings. The newly built areas, however, are usually filled with superblocks and wide roads as a result of car-prioritized development. In Chongqing, dense roads are only accessible in several commercial centers. When it comes Shijiazhuang, fine-scale blocks can only be found in very limited areas. The tradition of focusing

(a)

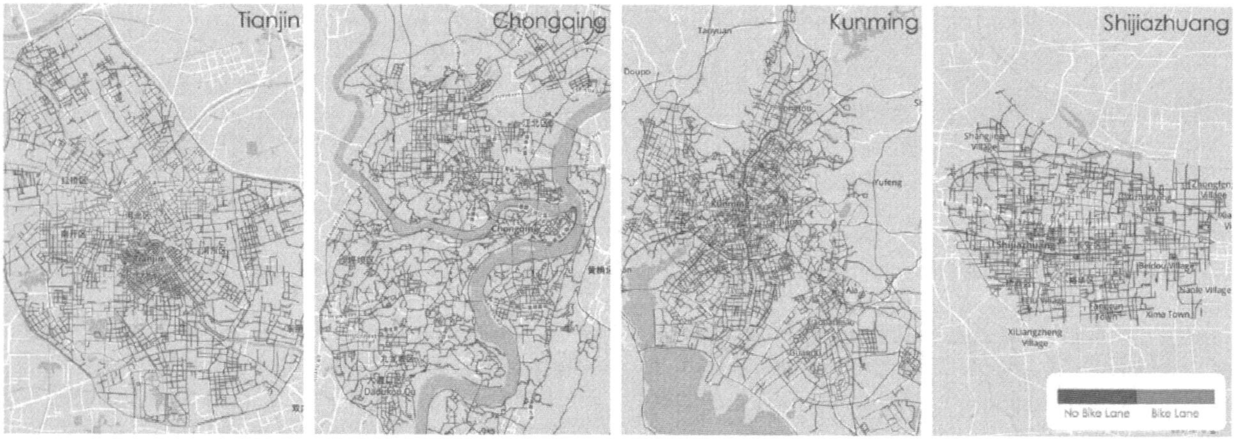

(b)

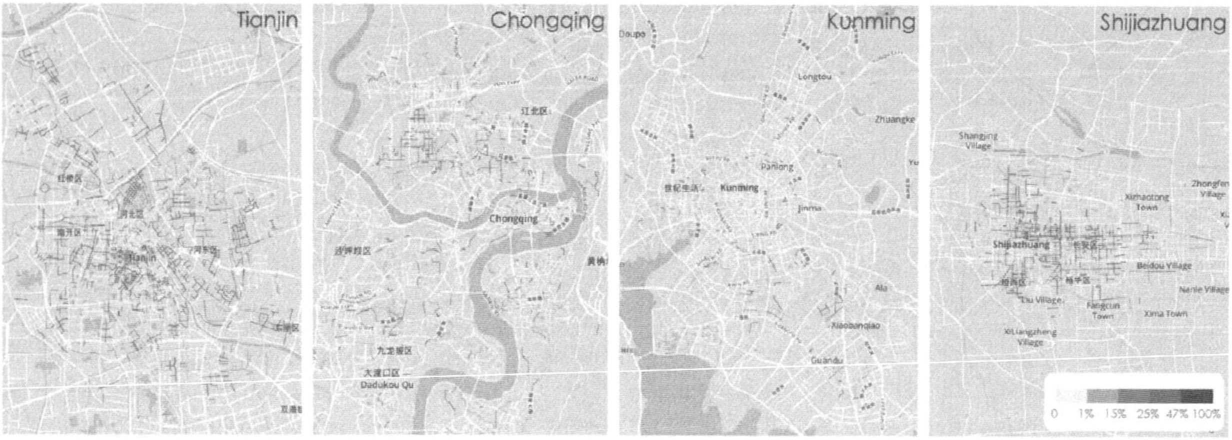

(c)

(continued)

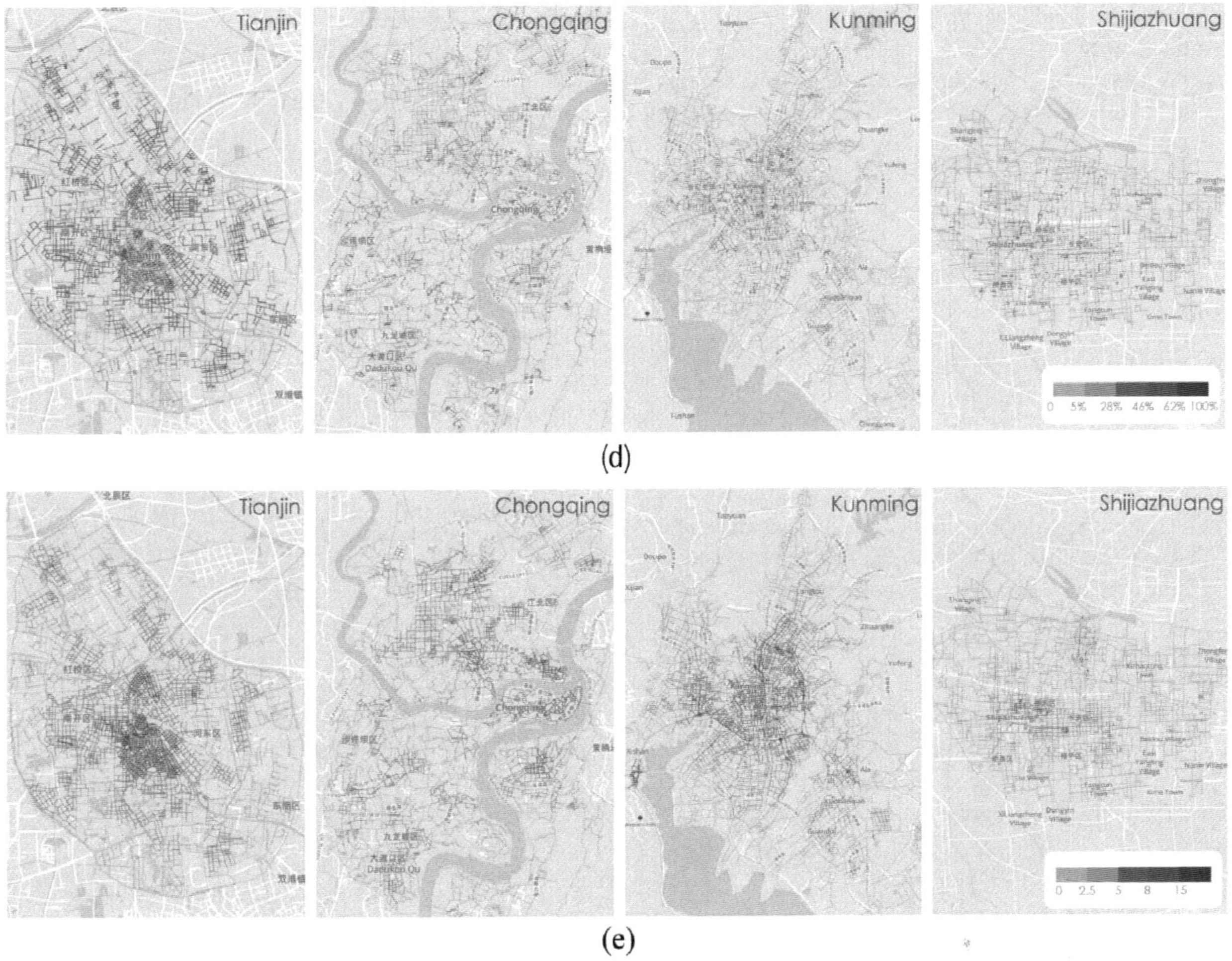

Figure 3. Street indictor map over four cities showing (*a*) sidewalks, (*b*) bike lanes, (*c*) percentage of sidewalks with illegal parking, (*d*) street wall continuity ratio, and (e) street network density (km/km^2).

on motorized vehicles is sure to be abandoned. Instead, urban planning should bring the right of the road back to the people by building pedestrian-friendly street networks.

Conclusion and Discussion

Extending existing scoring approaches, we took advantage of open source data and established a comprehensive and scalable street score framework for walkability and bikeability to evaluate urban streets in China. This method was applied to four Chinese cities: Tianjin, Chongqing, Kunming, and Shijiazhuang. The initial assessment results showed that these four cities have formed clusters of high-walkability streets in city centers, while in peripheral areas, street walkability fades out. A preliminary bikeable street network is evident in Tianjin, Kunming, and Shijiazhuang. However, the prevalence of bike lanes on streets is much lower than that of sidewalks. Regarding street sub-scores and indicators, the problem of illegal parking has increased, especially in prosperous city centers, which poses a danger for walking and biking safety. Biking safety and comfort are compromised due to a lack of physically separated bike lanes in all four cities. Furthermore, the street wall continuity ratio varies from place to place, and a few clustered areas exhibit a continuity of building facade along the street. As for the street network density, there are obvious gaps between traditional city centers and newly developed car-oriented areas. These findings could give us the following policy implications:

(1) Bikeability deserves more attention than walkability in Chinese cities. Our study has shown that the general biking environment is worse than the walking environment in China. More importantly, the recent bike-sharing boom in China has also triggered a boom in bike use, calling for policy interventions to meet this new demand. Adding physically separated

bike lanes to form a more fine-grained continuous network should be an effective way to promote bikeability.

(2) Policy makers should develop better parking enforcement measures to address the predominant illegal parking problem on streets. Improvement in city centers is more urgently called for. Otherwise, the safety of walking and biking in China will remain significantly undermined.

(3) Urban and transportation planners should advocate urban patterns of denser streets, smaller blocks, and narrower streets, especially in new urban towns, to enhance the overall comfort and convenience for walking and biking.

This paper has several limitations. First, this approach mainly depended on online open source data and street-view image detection. Results may be biased since street images were taken during different hours of the day and consistency across cities might not be guaranteed. Validation of this approach using observational data and/or other sources of data is recommended for future research. Second, the collinearity issue among individual indicators was not discussed in this paper. Finally, as data availability may be different in other countries, some metrics proposed in this study would require further customization for street score assessment, although the general approach should be adaptable.

Acknowledgments

This work is supported by the National Natural Science Foundation of China (No.51378278).

References

1. Ministry of Housing and Urban-Rural Development. *China's Urban Pedestrian and Bicycle Transportation System Planning and Design Technical Guideline*. Ministry of Housing and Urban-Rural Development, 2013 (in Chinese).
2. Shanghai Municipal Planning and Land & Resources Administration. *Shanghai Street Design Guideline*. Shanghai Municipal Planning and Land & Resources Administration, 2016 (in Chinese).
3. Osgood, C. E., G. J. Suci, and P. H. Tannenbaum. *The Measurement of Meaning*. University of Illinois Press, Illinois, 1964.
4. Gehl, J., and L. Gemzoe. *Public Space Public Life*. The Danish Architectural Press, Copenhagen, 1996.
5. Ratemystreet.com. http://www.ratemystreet.co.uk. Accessed December 1, 2013.
6. Rosenberg, D., D. Ding, J. F. Sallis, J. Kerr, G. J. Norman, N. Durant, S. K. Harris, and B. E. Saelens. Neighborhood Environment Walkability Scale for Youth (NEWS-Y): Reliability and Relationship with Physical Activity. *Preventive Medicine*, Vol. 49, No. 2, 2009, pp. 213–218.
7. Frank, L. D., T. L. Schmid, J. F. Sallis, J. Chapman, and B. E. Saelens. Linking Objectively Measured Physical Activity with Objectively Measured Urban Form: Findings from SMARTRAQ. *American Journal of Preventive Medicine*, Vol. 28, No. 2, 2005, pp. 117–125.
8. Frank, L. D., J. F. Sallis, T. L. Conway, J. E. Chapman, B. E. Saelens, and W. Bachman. Many Pathways from Land Use to Health: Associations Between Neighborhood Walkability and Active Transportation, Body Mass Index, and Air Quality. *Journal of the American Planning Association*, Vol. 72, No. 1, 2006, pp. 75–87.
9. Pikora, T., B. Giles-Corti, F. Bull, K. Jamrozik, and R. Donovan. Developing a Framework for Assessment of the Environmental Determinants of Walking and Cycling. *Social Science & Medicine*, Vol. 56, No. 8, 2003, pp. 1693–1703.
10. Winters, M., M. Brauer, E. M. Setton, and K. Teschke. Mapping Bikeability: A Spatial Tool to Support Sustainable Travel. *Environment and Planning B: Planning and Design*, Vol. 40, No. 5, 2013, pp. 865–883.
11. Lu, Y. Neighborhood Community Walkability Based on Life Community Service: Case Study of Jiangpu Road in Shanghai. *Urban Planning Forum*, Vol. 5, 2013, pp. 113–118 (in Chinese).
12. Wu, J., W. Qin, J. Peng, and W. Li. A Life Facility Evaluation Based on Walk Score, Talking Futian District in Shenzhen as an Example. *Urban Development Research*, 2014, pp. 49–56 (in Chinese).
13. Liu, J., G. Lin, and X. Lan. An Environment Evaluation for Walking Comfort in Anshandao Historic Neighborhood in Tianjin. *Urban Architecture*, 2015, pp. 121–122 (in Chinese).
14. Lin, L., and A. V. Moudon. Objective Versus Subjective Measures of the Built Environment, Which Are Most Effective in Capturing Associations with Walking? *Health & Place*, Vol. 16, No. 2, 2010, pp. 339–348.
15. Leslie, E., N. Coffee, L. Frank, N. Owen, A. Bauman, and G. Hugo. Walkability of Local Communities: Using Geographic Information Systems to Objectively Assess Relevant Environmental Attributes. *Health & Place*, Vol. 13, No. 1, 2007, pp. 111–122.
16. Owen, N., E. Cerin, E. Leslie, N. Coffee, L. D. Frank, A. E. Bauman, G. Hugo, B. E. Saelens, and J. F. Sallis. Neighborhood Walkability and the Walking Behavior of Australian Adults. *American Journal of Preventive Medicine*, Vol. 33, No. 5, 2007, pp. 387–395.
17. Carr, L. J., S. I. Dunsiger, and B. H. Marcus. Walk Score™ as a Global Estimate of Neighborhood Walkability. *American Journal of Preventive Medicine*, Vol. 39, No. 5, 2010, pp. 460–463.
18. Hajna, S., K. Dasgupta, M. Halparin, and N. A. Ross. Neighborhood Walkability: Field Validation of Geographic Information System Measures. *American Journal of Preventive Medicine*, Vol. 44, No. 6, 2013, pp. 55–59.
19. Van Dyck, D., B. Deforche, G. Cardon, and I. De Bourdeaudhuij. Neighbourhood Walkability and Its Particular Importance for Adults with a Preference for Passive Transport. *Health & Place*, Vol. 15, No. 2, 2009, pp. 496–504.
20. Millington, C., C. W. Thompson, D. Rowe, P. Aspinall, C. Fitzsimons, N. Nelson, and N. Mutrie, and SPARColl—The Scottish Physical Activity Research Collaboration. Development of the Scottish Walkability Assessment Tool (SWAT). *Health & Place*, Vol. 15, No. 2, 2009, pp. 474–481.
21. Front Seat. *Walk Score Methodology*, 2011. http://www.walkscore.com/methodology.shtml. Accessed June 7, 2011.

22. Yin, L., Q. Cheng, Z. Wang, and Z. Shao. 'Big Data' for Pedestrian Volume: Exploring the Use of Google Street View Images for Pedestrian Counts. *Applied Geography*, Vol. 63, 2015, pp. 337–345.
23. Yin, L., and Z. Wang. Measuring Visual Enclosure for Street Walkability: Using Machine Learning Algorithms and Google Street View Imagery. *Applied Geography*, Vol. 76, 2016, pp. 147–153.
24. Long, Y., and L. Liu. How Green Are the Streets? An Analysis for Central Areas of Chinese Cities Using Tencent Street View. *PLoS One*, Vol. 12, No. 2, 2017, p. e0171110.
25. Jiang, Y., P. Gu, Y. Chen, D. He, and Q. Mao. Influence of Land Use and Street Characteristics on Car Ownership and Use: Evidence from Jinan, China. *Transportation Research Part D: Transport and Environment*, Vol. 52, 2017, pp. 518–534.
26. Wang, D., Y. Lu, W. Zhu, and J. Zhang. Development and Application of a Computer Assisted Walkability Evaluation System at the Perspective of Daily Community Service Facilities. *Journal of Tongji University (Natural Science)*, Vol. 43, No. 12, 2015, pp. 1815–1822 (in Chinese).
27. Shannon, C. E., and W. Weaver. *The Mathematical Theory of Communication*. University of Illinois Press, Illinois, 1949.
28. Lopez-Garcia, P., E. Onieva, E. Osaba, A. D. Masegosa, and A. Perallos. GACE: A Meta-Heuristic Based in the Hybridization of Genetic Algorithms and Cross Entropy Methods for Continuous Optimization. *Expert Systems with Applications*, Vol. 55, 2016, pp. 508–519.
29. Selvan, S. E., M. S. P. Subathra, A. H. Christinal, and U. Amato. On the Benefits of Laplace Samples in Solving a Rare Event Problem Using Cross-Entropy Method. *Applied Mathematics and Computation*, Vol. 225, 2013, pp. 843–859.
30. Chen, F., J. Wang, and Y. Deng. Road Safety Risk Evaluation by Means of Improved Entropy TOPSIS-RSR. *Safety Science*, Vol. 79, 2015, pp. 39–54.
31. Mensi, W., M. Beljid, and S. Managi. Structural Breaks and the Time-Varying Levels of Weak-Form Efficiency in Crude Oil Markets: Evidence from the Hurst Exponent and Shannon Entropy Methods. *International Economics*, Vol. 140, 2014, pp. 89–106.
32. Ngoduy, D., and M. J. Maher. Calibration of Second Order Traffic Models Using Continuous Cross Entropy Method. *Transportation Research Part C: Emerging Technologies*, Vol. 24, 2012, pp. 102–121.
33. Tianjin Urban Planning and Design Institute. In 4th Tianjin Comprehensive Transport Survey, 2012, p. 9 (in Chinese).
34. Beijing Transportation Research Center. In *Beijing Transport Annual Report*, 2015, p. 26. (in Chinese)
35. Shanghai City Comprehensive Transportation Planning Institute. In 5th Shanghai Comprehensive Transport Survey, 2015, p. 23 (in Chinese).
36. Transport for London. *London Travel Demand Survey*, 2015.
37. The Seoul Institute. In *Seoul Statistical Series*, 2015, p. 76.
38. Cervero, R., and K. Kockelman. Travel Demand and the 3Ds: Density, Diversity, and Design. *Transportation Research Part D: Transport and Environment*, Vol. 2, No. 3, 1997, pp. 199–219.
39. Ewing, R., and R. Cervero. Travel and the Built Environment: A Meta-Analysis. *Journal of the American Planning Association*, Vol. 76, No. 3, 2010, pp. 265–294.
40. The State Council of the Communist Party of China. *Some Views on Further Strengthening the Management of Urban Planning and Construction*. The State Council of the Communist Party of China, 2016 (in Chinese).

The Standing Committee on Transportation in the Developing Countries (ABE90) peer-reviewed this paper (18-05964).

Article

TRR
JOURNAL OF THE TRANSPORTATION RESEARCH BOARD

Transportation Research Record
2018, Vol. 2672(31) 76–86

DOI: 10.1177/0361198118759075
journals.sagepub.com/home/trr
SAGE

Modeling Pedestrian Crossing Behavior and Safety at Signalized Intersections

S. Marisamynathan[1] and P. Vedagiri[1]

Abstract

Pedestrian noncompliance behavior is one of the most critical causes of pedestrian involved traffic crashes at intersections in India. Thus, the objectives of this study are to examine how various factors affect pedestrian crossing behavior and to propose models for pedestrian crossing behavior and level of safety at signalized intersections which will be useful to regulate pedestrian flow. The data were collected with video and a user perceptions survey at six selected signalized intersections in Mumbai, India. The differences between pedestrian crossing behavior with respect to personal characteristics, socioeconomic attributes, and existing crossing facilities were identified using Pearson correlation and odd ratio tests. Furthermore, the major reasons for noncompliance behavior were obtained by analysis of field data to prevent noncompliance behavior and enhance pedestrian safety. The results showed that a significant number of the pedestrians violated the traffic signal to save time and for convenience (46%). A binary logit model was developed to evaluate the impacts of contributing factors on pedestrian crossing behavior. Further, an ordered probability model was established to evaluate and estimate the pedestrian level of safety at signalized intersections. Two models were validated, and their statistical results show that the models predict the pedestrian crossing behavior and safety level more precisely. Developed models and study outcomes can help transport planners and designers understand pedestrian crossing behavior on crosswalks at signalized intersections and thus create a safer crossing environment for all pedestrians.

Traffic crashes involving pedestrians are one of the major causes of death and injury all over the world, particularly in developing countries. More than 270,000 pedestrians are killed in traffic crashes each year, which accounts for about 22% of total fatalities worldwide (*1*). The condition of pedestrian safety is even worse in developing countries such as India. In India, the crash rate between pedestrians and vehicles is relatively higher than the rates elsewhere in the world. According to the statistics, over 141,500 pedestrians died and 488,731 pedestrians were injured in road traffic crashes in the year 2014, which was more than the previous year (*2*). More than 60% and 50% of pedestrian crashes occurred at signalized intersections in Japan and China respectively, whereas in India nearly 65% of pedestrian crashes happened at signalized intersections. In India, pedestrian noncompliance behavior is frequent and thus has become a main factor in the many pedestrian crashes at signalized intersections.

Pedestrian crossing behavior highly depends on the surrounding environment and traffic situations at signalized intersections. The pedestrian crosswalk area and time available to cross the crosswalk are limited due to the vehicles and road infrastructure. In most signalized intersections, pedestrians are released with the parallel through movement and right-turn vehicles in a shared signal phase system, and there is free left-turning movement of vehicles which leads to pedestrian and vehicle conflicts in crosswalks. Therefore, pedestrians start to cross the crosswalk by noncompliance with the traffic signal in developing countries, especially in India. Another study shows that 55% of pedestrians were noncompliant with traffic signals at signalized intersections in the city of Mumbai, India, in 2016 (*3*). The decisions involved in pedestrian noncompliance are influenced by many factors such as personal characteristics, socioeconomic attributes, behavior of pedestrians and existing crossing facilities. In addition, while various existing studies have studied pedestrian noncompliance behavior, studies focusing on the topic in India and other developing countries are still important to prevent noncompliance behavior and enhance pedestrian safety. Analyzing pedestrian crossing behavior is challenging due to the complex and diverse nature of pedestrians. It is challenging to estimate accurate pedestrian safety level because of ranked and ordered qualitative data. Modeling pedestrian level of safety by combining quantitative and qualitative data is a very difficult task and missing in existing studies, but it can produce more accurate pedestrian

[1]Transportation Systems Engineering, Department of Civil Engineering, Indian Institute of Technology Bombay, Mumbai, India

Corresponding Author:
Address correspondence to S. Marisamynathan: marisamy1989@gmail.com

safety levels at signalized intersections. In light of the disappointing state of pedestrian safety, this study examines various influencing variables on pedestrian crossing behavior and develops suitable models to estimate pedestrian crossing behavior and safety level at signalized intersections. This study can be useful to understand pedestrian crossing behavior and to propose more effective countermeasures to reduce pedestrian involved crashes at signalized intersections in India.

Literature Review

Most of the existing studies have focused only on motorized vehicular traffic safety and very few studies have concentrated on non-motorized traffic safety. In the initial stage, existing studies focused on evaluating pedestrian crossing characteristics such as spatial and behavioral characteristics. Various studies have analyzed the influential factors of pedestrian crossing behavior with respect to socioeconomic, psychological, road and other traffic characteristics (*1, 4, 5*). Compared with drivers, pedestrian behavior is often unpredictable, because pedestrians are much freer to move or choose their path and they are not constrained by regulations. Therefore, a few studies have explored pedestrian movements with cameras and by conducting questionnaire surveys so that pedestrians' perceptions about crossing environments could be collected (*6*).

Many existing studies have examined the differences in pedestrian crossing behavior with respect to gender and age (*7, 8*). Mostly male pedestrians tend not to comply with traffic signals more often than female pedestrians and are more likely to cross the crosswalk in risky conditions (*7*). Female pedestrians were more sensitive to risk perception and gave less preference to crossing the crosswalk during the red-light phases. Adult pedestrians are more likely than older pedestrians to chose not to comply with traffic laws (*9*). Also, older pedestrians had a higher safety perception than younger pedestrians by using safer facilities, whereas the adult pedestrians more often focused on how to arrive at the destination more quickly (*10–12*).

Some studies have focused on pedestrian crossing speed and their attitude effect on crossing behaviors (*1, 13*). Pedestrian crossing speed is an important parameter in safety analysis, and plays a significant role in the geometric and signal design of intersections. Existing results found that pedestrian crossing speed of male pedestrians was higher than females, and with increased age, pedestrian crossing speed would slow down (*3*). Also, the crossing speed of pedestrians in groups tended to be slower than single pedestrians (*14*). In addition, the tendency to cross on the red phase was higher when fewer pedestrians were waiting at the curb.

Pedestrian crossing behavior is influenced by various elements of the road crossing environment such as signalized and unsignalized intersections, marked and unmarked crosswalk, with and without median, and pedestrian warning signs (*15*). The waiting time of pedestrians was also found in other studies to have a significant impact on noncompliance behavior (*16, 17*). Pedestrians are more likely to choose not to comply with traffic laws if the pedestrian signal waiting time is high (*18*). In addition, some studies found that pedestrians accompanied by children or heavy luggage tended to comply with traffic signals. Some studies have examined the usage and impact of mobile phones on pedestrian crossing behaviors (*19*). The results showed that pedestrians who used mobile phones while crossing the road were more likely to be involved in traffic crashes. Also, using a mobile phone was associated with slower crossing speeds of pedestrians at signalized intersections and it may undermine pedestrian safety (*19–21*).

Existing studies have examined many techniques to model pedestrian noncompliance behaviors. A hierarchically multivariate regression, structural equation, stepwise regression, bi-level multivariate, binomial logistic, hybrid framework, and Monte Carlo methods were used to examine pedestrian road crossing behavior, compliance with safety rules, unsafe crossing, crossing choice and pedestrian safety level in crosswalks at signliazed intersections (*4, 7, 17, 22, 23*).

Finally, existing studies have analyzed pedestrian noncompliance with traffic signals considering various factors such as psychological characteristics and socioeconomic, traffic, and road factors. However, existing studies have seldom considered the impact of pedestrian attitude related parameters on crossing behaviors at signalized intersections. Pedestrian crossing behaviors are affected by attitudes, hence it would be essential to consider such factors in the analysis. Furthermore, traffic conditions, signal system, crossing facilities, cultural background, and decision-making processes in India are quite different from those in other countries. Thus, it is very important to conduct research on pedestrian noncompliance behavior and conditions of safety on the crosswalk at signalized intersections in India, using an integrated approach that includes a questionnaire survey to obtain pedestrian perceptions.

Research Objectives

The objectives of this study are as follows: (*a*) to examine pedestrian noncomplaince behavior with traffic signals at signalized intersections based on pedestrian perception data; (*b*) to identify which variables significantly affect pedestrian crossing behavior and safety level, and compare the differential effects of these factors; (*c*) to develop a pedestrian crossing behavior model by considering various influencing variables at signalized intersections; and (*d*) to develop a pedestrian level of safety model at crosswalks of signalized intersections by combining qualitative and quantitative data.

Table 1. Study Location, Samples, and Basic Traffic Information

Intersection ID	Intersection name	Pedestrian signal phase (s)			Average pedestrian volume per approach (ped/h)	Number of respondents	Noncompliance rate (%)	Presence of crosswalk marking	Crosswalk length (m)
		Green	Red	Flashing green					
A	Rajguru Chowk	22	75	3	436	63	45.9	Yes	31.5
B	Haj House Chowk	10	120	4	148	40	15.4	Yes	26
C	Hedgewar Chowk	10	120	3	320	50	20.1	No	19
D	IITB Police Chowki	32	180	3	336	77	16.7	No	33
E	Sakkinakka Junction	26	120	4	280	36	35.7	Yes	25
F	Sant Tukaram Maharaj Chowk	34	189	3	229	40	34.9	No	23

Data Collection on Pedestrian Crossing Behavior

Study Sites

Basic information about the selected study sites and samples are presented in Table 1. Before finally selecting six signalized intersections, a reconnaissance survey was conducted and more than 20 intersections were selected to assess the suitability. To incorporate variation in roadways, traffic, and signal characteristics, six signalized intersections were selected from a list of all the signalized intersections in Mumbai suburban area, India. The locations were operated by fixed signal time and pedestrians, at the intersections, as per design, were supposed to cross in a shared signal phase system, with bi-directional pedestrian flow. Exclusive turning lane and channelizations were not provided at the study locations. Pedestrian volumes were also counted during peak hour. Basic traffic, pedestrian and geometric characteristics are presented in Table 1.

Data Collection Method

Pedestrian crossing behaviors were studied through the use of data collected by field observation of pedestrian crossings with a video camera set up next to crosswalks at each intersection. The crosswalks were selected at the major approach of intersections and the camera covered the pedestrians, traffic characteristics, and traffic signals. The questionnaire survey was also conducted simultaneously during video recording at the six selected signalized intersections. Data was collected on weekdays during evening peak hours (4:00 to 6:30 p.m.) and weather conditions were sunny and warm during survey duration.

Details of Collected Data

The video recording offered information about pedestrian crossing volumes, crossing behavior, crossing time, compliance behavior, crossing location, pedestrian signal phase time, platoon size, mobile phone usage while crossing, type of crossing (single stage or two-stage crossing) and personal characteristics (age, gender). The pedestrian crossing speeds acquired were average crossing speeds, defined as the crosswalk distance divided by the pedestrian crossing time. The mentioned data were extracted by data collectors from the video recording using AVS video editor software.

As the information on pedestrian age was extracted from the video recording, the data collectors could not define the exact age of any given pedestrian but, instead, estimated the age group to which the pedestrian likely belonged. To get more appropriate and realistic statistical data, pedestrian ages were divided into six groups: less than 12 years, 12 to 16, 19 to 25, 26 to 35, 36 to 55, and more than 55. The pedestrian compliance rate was calculated with the data extracted from the video recording. Pedestrian crossing during the pedestrian green-light phase is considered compliance with the signal. Thus, the pedestrian compliance rate at signalized intersection is defined as the number of pedestrian crossing the crosswalk during the pedestrian green phase out of the total number of pedestrians who crossed the crosswalk during the selected duration.

The type of activity which pedestrians were performing on their mobile phones, while crossing the crosswalk, was obtained from analysis of the video recording. These were divided into four groups: talking, listening to music, texting (or other form of visual usage) and not using. The details of pedestrian behavior with respect to traffic flow were extracted for given pedestrians from the video.

Questionnaire Survey for Pedestrian Perception Data

To understand pedestrians' perceptions on crossing behavior and safety level, a questionnaire survey was conducted in the field. Based on a literature review and field conditions, a questionnaire form was prepared which could be used to develop pedestrian crossing behavior and level of safety models. The questionnaire used for the survey consisted of 13 questions about what pedestrians do and how they feel when they cross at crosswalks of signalized intersections. The questions covered the following aspects: (*a*) demographic information (gender, age, education level, occupation, and trip purpose), (*b*) pedestrian crossing choice at particular crosswalk of signalized intersections (how frequently they use the selected crosswalk, how frequently they fail to comply with the signal in a week, tolerable waiting time, and crossing type such as running or walking), (*c*) pedestrian perceptions on safety related to traffic conditions, crosswalk facilities and signal system, and (d) the major reason for pedestrian noncompliance with traffic signal.

Questionnaire data was collected at the selected study sites by using several qualified enumerators, at the same time as the field was being video recorded. Enumerators stood on both sides of the crosswalks and questions were asked to pedestrians immediately after they crossed the crosswalk, to give the crossing experience in realistic site conditions. To collect details about noncompliance behavior, the questionnaire survey was conducted with pedestrians who were noncompliant with the signal. A total of 306 pedestrians responded to the questionnaire survey and the number of pedestrians surveyed in each location are presented in Table 1. The detailed information for all respondents is explained in the section on reasons for pedestrian noncompliance behavior.

Pedestrian Crossing Patterns and Noncompliant Behavior in Crosswalks

The collected video and questionnaire data provided the following information on pedestrians: gender, age group, education level, occupation, trip purpose, crosswalk usage, tolerance of waiting time, crossing behavior, and the compliance behavior with the signal. A total of 306 pedestrians participated in the questionnaire survey to obtain complete information in the field. Descriptive statistics of the pedestrians' characteristics and behaviors are listed in Table 2.

From the collected data, the proportion of male pedestrians is greater than the proportion of female pedestrians. The data show that pedestrians from the age groups of 19–25 years and 26–35 years formed the majority of those who responded to the questionnaire survey. Also, the proportion of pedestrians with undergraduate education is the highest compared with the proportions of pedestrians with other education levels. For pedestrian occupation, the majority were either student and private employee. The purpose of the pedestrians' trip was going to work in just over half of cases, with the other half relatively evenly distributed between: shopping, recreation, education, other purpose and social activities. More than 50% of pedestrians used the crosswalk regularly and more than 60% stated that they did not comply with the signal either occasionally or regularly.

From the observed data, only 22% of pedestrians tolerated more than 80 seconds waiting time and others chose not to comply with the signal due to more tolerance waiting time. The noncompliance rates in locations A, B, C, D, E, and F are 45.9, 15.4, 20, 16.7, 35.7, and 30% respectively. These data show that A ranks first, followed by E, F, C, D and B. Location A has the largest noncompliance rate because it has a high pedestrian flow, shorter pedestrian green-light phase, and pedestrians may not understand the traffic regulations or signal phase system. Location B has the mixed land-use pattern (Institute, Government Sector, Business Center and Private Offices), and so the pedestrians at location B may have a relatively greater sense of responsibility to obey the traffic laws than those at intersections with other land-use patterns. From the observed data, 69% of pedestrians crossed the crosswalk in a single-stage crossing and others crossed in a two-stage crossing by utilizing the median width or refuge islands.

Reasons for Pedestrian Noncompliance Behavior

Of the 306 individuals who responded to the questionnaire survey, 68% were male and the average speed of the respondent was 1.21 m/s. In the questionnaire survey, pedestrians were asked to answer questions about their personal and socioeconomic characteristics and experience with noncompliance at signalized intersections. Further, pedestrians who chose not to comply with the traffic signal were offered eight reasons and asked to choose the one that best represents the reason why they crossed the crosswalk during the red-light phase at a signalized intersection. The eight reasons were listed based on the literature review and the field conditions of selected sites. The results are shown in Table 3.

The major reason for pedestrian noncompliance was the desire to save time and for convenience (46.7%). Another major reason was less or no traffic flow at particular approach and there is no risk (29.1%) if there is noncompliance with the signal. Further, 8.2% of the pedestrians responded that they chose not to comply with the signal because the signal system and timing were not properly established: either the pedestrian red phase is too long or the pedestrian green time is too short. In addition, other factors, such as because other people do this, no accidents will happen if I do this one or two times, and no traffic police management, account for more than 6.5%. More than 4% of the pedestrians answered that they chose not to comply with traffic signal mainly because the facilities are not properly provided: absence of crosswalk marking or refuge island or waiting area, crosswalk conditions, and no connection between sidewalk and crosswalk.

Pedestrian Crossing Behavior at Crosswalks

Factors Affecting Pedestrian Crossing Behavior at Crosswalks

To understand whether demographic or social parameters or existing facilities have an impact on pedestrian crossing

Table 2. Descriptive Statistics of Pedestrian Characteristics and Behaviors

Pedestrian characteristic/behavior	Percentage
Gender	
Male	68.1
Female	31.9
Age group	
Less than 12	0
12–18 years	3.6
19–25 years	35.1
26–35 years	32.7
36–55 years	21.9
More than 55 years	6.8
Education level	
Primary school (1–5)	2.4
Middle school (6–10)	10
Senior school (11–12)	17.9
Undergraduate	43.4
Postgraduate	25.9
Occupation	
Student	31.5
Laborer/daily worker	7.2
Private employee	28.3
Government employee	10.8
Business	11.2
Retiree	3.6
Others	7.6
Trip purpose	
Work	51.4
Shopping	12.7
Recreational	10.4
Education	10
Social	6
Others	9.6
Frequency of using the selected crosswalk	
Regularly (daily)	52.6
Rarely	37.8
Today only	9.6
Frequency of noncompliance with signal in a week	
Never	34.7
Occasionally	47.4
Always	17.9
Tolerable waiting time (waiting time acceptable for pedestrian to comply with signal)	
Less than 20 s	24.7
21–40 s	26.3
41–60 s	19.9
61–80 s	6.4
More than 80 s	22.7
How pedestrian crossed the particular crosswalk	
Above normal speed/running	10
Walking	60.2
Both	29.9

(Continued)

Table 2. Continued

Pedestrian characteristic/behavior	Percentage
Mobile phone usage while crossing	
Talking	6.6
Listening to music	5.6
Texting	3.4
Not using	84.4

patterns and behavior, the Pearson correlation test was first applied in SPSS software. If the Pearson sign value (p) is greater than 0.05, the observation is consistent with the null hypothesis, namely, that no significant differences in choices exist between two groups.

For further analysis, each variable was analyzed by use of a single-factor model with odd ratio (OR) statistics. The OR denotes the odds that an outcome will occur given a particular exposure, compared with the odds of the outcome happening in the absence of that exposure. For instance, the number of exposed cases and non-cases for crosswalk usage are a and b respectively, and the number of unexposed cases and non-cases for crosswalk usage are c and d, respectively, then the OR value is calculated as follows:

$$OR = \frac{a / c}{b / d} = \frac{ad}{bc} \tag{1}$$

If the OR is greater than 1, the exposure associated with higher odds of the outcome and the particular group is more likely to comply with signal; otherwise, the group may show a lesser tendency to comply with signal. It is used to examine the effects of age, education level, occupation, frequency of crosswalk usage, trip purpose and existing facilities on crossing patterns and behavior.

The analysis in this paper focuses on two types of crossing behaviors: compliance or noncompliance with traffic signal. Socioeconomic status is well known to be an influencing factor for pedestrian crossing behavior on the crosswalk. Based on Pearson correlation and OR test results, pedestrian education level, occupation, and trip purpose have a significant effect on pedestrian crossing behavior. In addition, the study identified that existing signal system and crossing facilities also have an impact on pedestrian crossing behavior at signalized intersections. All these factors were considered as independent variables and analyzed with pedestrian crossing behaviors.

Pedestrian Education Level. The crossing behavior of pedestrians with undergraduate-level education was most compliant (63.8%), followed by pedestrians with postgraduate (21.7%), senior school (21.1%) and middle school education level (11.8%). As expected, the least compliant were pedestrians with primary education level (4.1%). Therefore, higher levels of education (having a university degree or more) was

Table 3. Major Reasons for Pedestrian Noncompliance with Traffic Signal

Sl No	Reason	Number of respondents	Percentage
1	To save time (or in a hurry)	143	46.7
2	Because other people do this (following others or leader pedestrians)	9	2.9
3	No accidents will happen if I do this one or two times	5	1.6
4	Less or no traffic flow and there is no risk	89	29.1
5	Absence of traffic police personnel	6	2.0
6	Not noticing the traffic signal	16	5.2
7	Issues in existing signal design (more waiting time or shared signal phase system)	25	8.2
8	Non-availability of pedestrian crossing facilities at intersections (absence of crosswalk marking/refuge island, crosswalk conditions, no connection between sidewalk and crosswalk)	13	4.2

found to be a key determinant of crossing behavior of pedestrians at signalized intersections.

As a result, the data from the Pearson correlation indicates that significant differences in pedestrian crossing behavior exist across various pedestrian education levels ($p = 0.010$, which is <0.05). For further calculations of the ORs, the results are taken into account to determine whether a pedestrian with lower or higher level of education is more likely to choose not to comply with traffic rules. The OR values for all groups of education level are more than one (middle school = 1.94, senior school = 2.26, graduate = 3.59, and postgraduate = 4.85), which suggests that individuals in all these groups tend to comply with signal. The study found that there was no major difference between the pedestrians educated up to degree (undergraduate) and educated above degree level (postgraduate), indicating that these two groups of pedestrians were similar in the pedestrian compliance behavior.

Pedestrian Occupation. The pedestrian occupation groups considered were: student, laborer/daily worker, private employee, government employee, business, retiree, and other. The frequency distributions of the noncompliance rate of each group were 24.3, 12.2, 27.0, 12.2, 9.5, 0.0, and 14.9% respectively. The sample of retirees was very small and hence cannot be used directly for further analysis. Hence, for further analysis, data for pedestrians classified as retiree was combined with those for pedestrians classified as other. The noncompliance rate of pedestrians with private employment was found to be higher than for other categories of occupation. The percentage distribution of pedestrians with government, business and other employments were found to be nearly the same, these subcategories were influenced the compliance behavior of pedestrians

When statistically tested, it was found that pedestrian occupations were significant, differing at 5% of level of significance with pedestrian crossing behavior ($p = 0.000$, which is <0.05). Further, OR values for all occupation types (for student, 0.39; for daily laborers, 0.47; for private employees, 0.46; for government employees, 0.22; and for others, 0.13) were less than one with respect to student, which indicates that individuals in all groups were more inclined to comply with traffic laws. The study found that all the nine pedestrians observed who retirees followed the traffic laws.

Pedestrian Trip Purpose. Based on the different purposes for which trips are made predominantly by the pedestrians in the study area, the following trip purposes were considered for study: work, shopping, recreational, education, social and other activities. The percentage distribution of these six cases with respect to noncompliance behavior was 54.1, 17.6, 8.1, 5.4, 8.1, and 6.8% respectively.

The Pearson correlation test was adopted to analyze whether differences in pedestrian crossing behavior for the six trip purposes are significant. The result is as follows: for noncompliant behavior, $p = 0.027$, which is <0.05 and it was observed that trip purpose did influence pedestrian compliance behavior. The noncompliant behavior of pedestrians whose purpose was "work" was higher than those with shopping, recreational, or social purposes, as pedestrians with work purposes tend to use the same route every day and are thus more familiar with the walking environment than those with other purposes. It can also be observed that the noncompliance rate of the pedestrians with recreational and social purposes was found to be nearly the same.

Existing Signal System and Crossing Facilities. The green-light phase times for pedestrians at the study area ranged from 10 to 34 seconds, but they were mainly concentrated within the interval of 20 to 30 seconds. Significant differences in compliance behavior with signal ($p = 0.007$, which is <0.05) exist between the existing signal systems. The results found that the longer the pedestrian red time, the more likely were pedestrians to disobey the traffic signal, which might be because of impatience or hurry or more waiting time. Due to a longer red time, more than 65% of pedestrians were found to be in noncompliance with traffic signal in the study area.

The results of the Pearson correlation show that significant differences in compliance behavior ($p = 0.042$, which is <0.05) exist with respect to non-availability of pedestrian crossing facilities. The OR value for non-availability of crossing facilities (1.11) exceeds one, which indicates that individuals in this group are more inclined to chose not to comply with traffic signal than when there are proper existing facilities. The absence of pedestrian crossing facilities such as crosswalk markings, median width and waiting area tend to increase noncompliance with traffic laws while crossing signalized intersections. Significant factors which have been identified in this analysis are used to develop the pedestrian crossing behavior model in the following section.

Model Pedestrian Crossing Behavior at Signalized Intersection Crosswalks

Generally, two types of crossing behavior, compliance and noncompliance, were observed when pedestrians crossed the crosswalk at signalized intersections. A logit model is used to predict a binary response from a binary predictor, used to predict the categorical dependent variable based on predictor variables. According to the pedestrian crossing behavior, there are only two possible outcomes: compliance or noncompliance while crossing the crosswalk. Therefore, a binary logit model was considered, using one if pedestrians were noncompliant or zero otherwise. Therefore, a binary logit model was defined as follows:

$$P(y_n) = \frac{e^X}{1+e^X} \tag{2}$$

$$X = \beta_0 + \beta_1 x_1 + \beta_2 x_2 + \ldots + \beta_N x_N \tag{3}$$

where $P(y_n)$ = probability of pedestrians' n compliance on the crosswalk, x = variables that determine the probability of a discrete outcome for n, β_N = estimating parameters, N = number of independent parameters and X = linear function of multiple explanatory variables x.

A binary logit (BL) model was developed using four significant parameters which were identified in the previous section and the model developed can be used to predict the probability of pedestrian noncompliance behavior in the crosswalk. A BL model was developed in NLOGIT software using 80% of pedestrian samples. The developed utility function is shown in Equation 4.

$$X = g(x) = -1.282 - 0.302x_1 + 0.331x_2 + 0.306x_3 + 0.886x_4 \tag{4}$$

where x_1 = pedestrian education level, x_2 = pedestrian occupation, x_3 = non-availability of pedestrian crossing facilities, and x_4 = noncompliance to save time. The explanatory variables were defined in Table 2. From Equation 4, variables were considered with a t-value of ±1.96 or higher at the 95 percentile confidence interval. The pseudo R-squared value and success prediction rate, which are widely used to test the logit model's goodness of fit, are considered to decide the overall model prediction. The statistical results show that success prediction rate and pseudo R-squared of the model are 62.50% and 0.289 respectively. The value of pseudo R-squared defines an acceptable relationship of 28.90% between the predictors and prediction with field data. The statistical results prove that the developed model has better prediction capabilities for pedestrian crossing behavior in crosswalks at signalized intersections.

From the collected data, 80% of training data were utilized for development of the pedestrian crossing behavior model and the remaining 20% data were utilized for validating the proposed model. The overall success prediction of the pedestrian crossing behavior model was observed as 65%. The value proves that the developed BL model has predictive capabilities for pedestrian crossing behavior at signalized intersection. The developed model can be used to understand pedestrian crossing behavior based on an objective to prevent noncompliance behavior and enhance pedestrian safety at signalized intersections.

Pedestrian Level of Safety in Crosswalks

Factors Affecting Pedestrian Level of Safety at Crosswalks

A five-point category scale (1 = very bad, 2 = bad, 3 = average, 4 = good, and 5 = very good) was adopted to evaluate pedestrians' level of safety in existing facilities at signalized intersections. Three major questions were asked during the pedestrian questionnaire survey, as follows:

1. What you think about the effect of traffic conditions with respect to safety at selected crosswalks?
2. What you think of the safety level of the crosswalk based on facilities provided to pedestrians at signalized intersections?
3. How would you like to rate the operation of the signal system (waiting time, presence of free left turn for vehicles, and shared signal phase for pedestrian green)?

From the questionnaire survey, the average score for the question about the traffic conditions at crosswalks is 2.42, which is between bad and average and indicates that pedestrian safely level is highly dependent on traffic status and that the pedestrian crossing environment is unsatisfactory in Mumbai. The average score for the question about the conditions of pedestrian crossing facilities is 2.76, which is between average and bad. The result indicates that the design standards and maintenance levels of pedestrian crossing facilities are inadequate in Mumbai. The average score with respect to signal timing (2.78) indicates that the pedestrians were not satisfied with signal time and they failed to comply with the signal even when they were aware of the danger. Therefore, signal timing should be improved as per the pedestrian perspective.

Many different parameters affect pedestrians' level of safety while crossing crosswalks. The Pearson correlation test was used to analyze the influence of traffic characteristics, signal timing, pedestrian behavior, and existing facilities with pedestrian level of safety. The result shows that roadway width (or crosswalk length), presence of free left-turning movement of vehicles, pedestrian noncompliance behavior, pedestrian waiting time, and median width had significant effects on pedestrian safety level with a score at 95% confidence interval. Significant factors are used to develop a pedestrian level of safety model and the details are explained in the next section.

Model Pedestrian Perceptions Based Level of Safety at Signalized Intersection Crosswalks

The purpose of this model is to understand the pedestrian safety level and to identify the various influencing parameters on pedestrian safety at signalized intersections. Hence, the overall objective is to evaluate the safety of pedestrians while crossing the crosswalk. Pedestrian characteristics and behavior, traffic characteristics, and crossing facilities were considered and other intersection parameters were not considered in the model. The conventional regression model developed by earlier researchers does not consider the effect of ordering of the dependent variables. In a conventional regression model, the observations are assumed to follow one particular probability distribution and the means of these particular probability distributions vary in some systematic schemes with the values of the independent variables. Therefore, the conventional regression model failed to develop models using ordered and discrete data. This study has adopted the ordered probability model (OPM) which is more suitable for discrete and ordered data. For estimating pedestrian level of safety, OPM provides an appropriate method for capturing the accurate category toward safety level with significant independent parameters. The pedestrian level of safety was categorized into five levels: category 1 is very bad, category 2 is bad, category 3 is average, category 4 is good, and category 5 is very good. The generalized form of OPM for a single latent variable (y*) is as follows:

$$y_i^* = \beta X_i + \varepsilon_i \tag{5}$$

where X is a vector of independent variables that determines the pedestrian level of safety, β is a vector of estimable parameters, and ε is a randomly disturbance term, assumed to follow a standard normal distribution N [0,1]. Using Equation 5, and removing the subscript i to simplify the notation, observed categories of pedestrian level of safety for each observation are defined as (numerical scale 1 to 5)

$$y = 1 \, if \; z \le 0 \tag{6a}$$

$$y = 2 \, if \; 0 < z \le \mu_1 \tag{6b}$$

$$y = 3 \, if \; \mu_1 < z \le \mu_2 \tag{6c}$$

$$y = 4 \, if \; \mu_2 < z \le \mu_3 \tag{6d}$$

$$y = 5 \, if \; z \ge \mu_3 \tag{6e}$$

where μ are threshold parameters that categorize the observed variable y (pedestrian level of safety) and μ is estimated jointly with β. If the ε is assumed to be a normal distribution with mean 0 and variance 1, then an OPM results with the selection probabilities:

$$P(y=1) = F(0-\beta X) \tag{7a}$$

$$P(y=2) = F(\mu_1-\beta X) - F(0-\beta X) \tag{7b}$$

$$P(y=3) = F(\mu_2-\beta X) - F(\mu_1-\beta X) \tag{7c}$$

$$P(y=4) = F(\mu_3-\beta X) - F(\mu_2-\beta X) \tag{7d}$$

$$P(y=5) = 1 - F(\mu_3-\beta X) \tag{7e}$$

Therefore the probability that y_i falls into the j th category is given by

$$P(y_i = j) = F(\mu_i - \beta X_i) - F(\mu_{j-1} - \beta X_i); j = 1,2,\ldots J \tag{8}$$

where F (.) is the cumulative distribution functions of standard normal distribution, and μ and μ_{j-1} are the upper and lower threshold values for the category J.

From the questionnaire survey, a sample consisting of 306 pedestrians was collected and the percentage of pedestrian level of safety in each category was: 11.2, 28.8, 44.8, 13.6 and 1.6%. Category 3 (average) was perceived by the greatest number of pedestrians and category 5 (very good) by the lowest number. The results indicate that most of the pedestrians experienced an average level of safety while crossing the crosswalk. In the previous section, the effect of each variable from the questionnaire survey was analyzed and significant variables were identified by applying the Pearson correlation test in SPSS software at 95% confidence interval. Using 80% of the pedestrian data, an ordered probability model was developed in NLOGIT 4.0 software by combining qualitative and quantitative data. In the model, categories of pedestrian level of safety were considered as the dependent variable and other significant parameters were considered as independent variables. Three threshold values were estimated from the

Table 4. Ordered Probability Model Results of Pedestrian Safety Level

Variables	Estimated coefficient	Standard error	*t*-value	*p*-value
Constant	7.189	1.569	4.583	0.000
Roadway width (or crosswalk length) (in m)	0.754	0.289	2.607	0.009
Presence of free left turning for vehicle	−2.308	1.000	−2.309	0.021
Pedestrian noncompliance behavior	0.270	0.143	2.888	0.029
Pedestrian waiting time (in s)	−0.031	0.007	−4.388	0.000
Median width (in m)	−10.877	4.003	−2.717	0.007
Threshold 1	1.092	0.088	12.478	0.000
Threshold 2	2.570	0.110	23.289	0.000
Threshold 3	3.803	0.213	17.832	0.000

observed five categories and the developed model results are presented in Table 4.

From Table 4, the *t*-value and *p*-value for each variable were significant at 95% confidence interval. A positive sign denotes that the explanatory variable increases the probability of a pedestrian accepting a higher category of level of safety with an increase in its magnitude and a negative sign indicates that an increase in the variable gives a probability of a pedestrian accepting a lower category of safety level with a decrease in its magnitude. A negative sign indicates that a decrease in the magnitude of the variables decreases the probability of a pedestrian accepting the safety level of a specific category. The goodness of fit of the developed model by McFadden's Pseudo R squared value is found to be 0.386 and the estimated success prediction value is 51.6%. From the *t*-statistic and *p*-value of each variable and the model goodness of fit, it can be concluded that the model estimates the pedestrian's tolerable waiting time quite well in mixed traffic conditions.

Model development used 80% of the data and the remaining 20% of the data was used to validate the reliability of the proposed pedestrian level of safety model. The statistical performance level of the proposed model was analyzed using Origin Pro 9 software. Mean absolute percentage error (MAPE), root mean square error (RMSE) and success prediction rate were 3.88%, 1.930%, and 47.5% respectively. Finally, the performance results of proposed models provide more precise and reliable solutions to estimate the pedestrian level of safety which is based on pedestrian perceptions of crosswalks at signalized intersections.

Conclusions

A high proportion of road crashes involve pedestrians, and most of the crashes are reported at signalized intersections. Pedestrian noncompliance behaviors are common, which leads to pedestrian crashes. To obtain more detailed information about pedestrian noncompliance behavior, this research studied the behavioral characteristics of pedestrians and the factors that influence their crossing behavior at signalized intersections. The required data were obtained from video survey and a user perceptions survey at six signalized intersections in Mumbai, India. Factors were analyzed by performing Pearson correlation and OR statistics and also two models: a pedestrian noncompliance behavior model and a pedestrian level of safety model, were developed to understand pedestrians' crossing behavior and existing facilities.

The results show that more than 60% of pedestrians did not comply with signals either occasionally or regularly. However, most of the pedestrians preferred to walk normally when they crossed the crosswalk.

The following conclusions are drawn based on statistical tests:

1. Higher levels of education were found to be key determinants of pedestrian compliance behavior at signalized intersections.
2. Pedestrians whose occupation was private employment were found to have more noncompliant behavior than those of any other occupation.
3. Pedestrians whose purpose was working were found to have more noncompliant behavior than those with other trip purposes.
4. The noncompliance rate increases as the pedestrian red-light phase time increases, whereas the noncompliance rate decreases as the waiting time decreases.
5. The length of a crosswalk was correlated to pedestrian crossing behavior and safety level. The noncompliance rate is lower and safety level is higher when the crosswalk is shorter.
6. About one third (31%) of pedestrians crossed the crosswalk in a two-stage crossing and narrower width of median influenced the pedestrian safety level. As median width decreases, pedestrians have to wait in the road area which ultimately affects pedestrian level of safety.
7. Pedestrians' level of safety was influenced when the intersection allowed free left-turning movement for vehicles.

The analysis results of questionnaire survey show that 46% of pedestrians were noncompliant with signals due to saving time (or in a hurry). Also, 30% of pedestrians attributed their noncompliant behavior to less or no traffic flow and when they were aware that there was no risk fromnon-compliance with the signal. In addition, the pedestrians were not satisfied with traffic conditions, crossing facilities, and signal timing and they rated level of safety score as 2.6 (between bad and average). Presence of crosswalk markings would reduce pedestrian noncompliance behavior. Noncompliance was found to be high among the pedestrians when the red-light phase is longer due to impatience caused by various factors. This study suggests that providing an optimal green time for pedestrians could decrease the rate of noncompliance at signalized intersections. Long crosswalks make it difficult for the pedestrian to cross in the stipulated time at normal crossing speed. In addition to this, absence of refuge-islands obliges the pedestrians to run and fail to comply with the signal during the red time. Therefore, there is a need to improve the pedestrian crossing environment which can ultimately improve pedestrian level of safety.

This study contributes in identifying the influencing variables for pedestrian noncompliance behavior and a systematic procedure to estimate pedestrian crossing behavior at signalized intersections. An alternative approach and framework of combining user perception and quantitative data is proposed in the development of the pedestrian level of safety model, which is fit for Indian mixed traffic conditions. This study helps traffic planners, designers, and policy makers understand pedestrian crossing behaviors at the crosswalks of signalized intersections. The findings of this study will be useful for reducing pedestrian noncompliance behavior and improving crossing facility design.

Acknowledgments

The data collection process was funded by John Hopkins International Injury Research Unit, Baltimore, USA. The authors also thank the interviewers and the voluntary respondents for their assistance with the data collection in the survey.

References

1. Zhang, W., K. Wang, L. Wang, Z. Feng, and Y. Du. Exploring Factors Affecting Pedestrians' Red-Light Running Behaviors at Intersections in China. *Accident Analysis and Prevention*, Vol. 96, 2016, pp. 71–78.
2. National Crime Records Bureau. *Accidental Deaths and Suicides in India*. Ministry of Home Affairs, National Crime Records Bureau, New Delhi, 2015.
3. Marisamynathan, S., and P. Vedagiri. Modeling Pedestrian Level of Service at Signalized Intersection Under Mixed Traffic Conditions. *Transportation Research Record: Journal of the Transportation Research Board*, 2017. 2634: 1–18.
4. Figliozzi, M. A., and C. Tipagornwong. Pedestrian Crosswalk Law: A Study of Traffic and Trajectory Factors that Affect Non-Compliance and Stopping Distance. *Accident Analysis & Prevention*, Vol. 96, 2016, pp. 169–179.
5. King, M. J., D. Soole, and A. Ghafourian. Illegal Pedestrian Crossing at Signalized Intersections: Incidence and Relative Risk. *Accident Analysis and Prevention*, Vol. 41, No. 3, 2009, pp. 485–490.
6. Ren, G., Z. Zhou, W. Wang, Y. Zhang, and W. Wang. Crossing Behaviors of Pedestrians at Signalized Intersections. *Transportation Research Record: Journal of the Transportation Research Board*, 2011. 65–73.
7. Rosenbloom, T. Crossing at a Red Light: Behaviour of Individuals and Groups. *Transportation Research Part F: Traffic Psychology and Behaviour*, Vol. 12, No. 5, 2009, pp. 389–394.
8. Lipovac, K., M. Vujanic, B. Maric, and M. Nesic. Pedestrian Behavior at Signalized Pedestrian Crossings. *Journal of Transportation Engineering*, Vol. 139, No. 2, 2013, pp. 165–172.
9. Holland, C., and R. Hill. Gender Differences in Factors Predicting Unsafe Crossing Decisions in Adult Pedestrians Across the Lifespan: A Simulation Study. *Accident Analysis and Prevention*, Vol. 42, No. 4, 2010, pp. 1097–1106.
10. Holland, C., and R. Hill. The Effect of Age, Gender and Driver Status on Pedestrians' Intentions to Cross the Road in Risky Situations. *Accident Analysis and Prevention*, Vol. 39, No. 2, 2007, pp. 224–237.
11. Tom, A., and M.-A. Granié. Gender Differences in Pedestrian Rule Compliance and Visual Search at Signalized and Unsignalized Crossroads. *Accident Analysis and Prevention*, Vol. 43, No. 5, 2011, pp. 1794–1801.
12. Granié, M.-A. Effects of Gender, Sex-Stereotype Conformity, Age and Internalization on Risk-Taking Among Adolescent Pedestrians. *Safety Science*, Vol. 47, No. 9, 2009, pp. 1277–1283.
13. Marisamynathan, S., and P. Vedagiri. A New Approach to Estimate Pedestrian Delay at Signalized Intersections. *Transport*, 2016, 33(1), pp. 249–259.
14. Gates, T. J., D. A. Noyce, A. R. Bill, and N. Van Ee. *Recommended Walking Speeds for Pedestrian Clearance Timing Based on Pedestrian Characteristics*. Presented at 85th Annual Meeting of Transportation Research Record, Washington, D.C., 2006.
15. Sisiopiku, V., and D. Akin. Pedestrian Behaviors at and Perceptions Towards Various Pedestrian Facilities: An Examination Based on Observation and Survey Data. *Transportation Research Part F: Traffic Psychology and Behaviour*, Vol. 6, No. 4, 2003, pp. 249–274.
16. Li, B. A Model of Pedestrians' Intended Waiting Times for Street Crossings at Signalized Intersections. *Transportation Research Part B: Methodological*, Vol. 51, 2013, pp. 17–28.
17. Hamed, M. M. Analysis of Pedestrians' Behavior at Pedestrian Crossings. *Safety Science*, Vol. 38, 2001, pp. 63–82.
18. Tiwari, G., S. Bangdiwala, A. Saraswat, and S. Gaurav. Survival Analysis: Pedestrian Risk Exposure at Signalized Intersections. *Transportation Research Part F: Traffic Psychology and Behaviour*, Vol. 10, No. 2, 2007, pp. 77–89.
19. Hatfield, J., and S. Murphy. The Effects of Mobile Phone Use on Pedestrian Crossing Behaviour at Signalized and Unsignalized Intersections. *Accident Analysis and Prevention*, Vol. 39, No. 1, 2007, pp. 197–205.

20. Schwebel, D. C., D. Stavrinos, K. W. Byington, T. Davis, E. E. O'Neal, and D. de Jong. Distraction and Pedestrian Safety: How Talking on the Phone, Texting, and Listening to Music Impact Crossing the Street. *Accident Analysis and Prevention*, Vol. 45, 2012, pp. 266–271.
21. Nasar, J., P. Hecht, and R. Wener. Mobile Telephones, Distracted Attention, and Pedestrian Safety. *Accident Analysis and Prevention*, Vol. 40, No. 1, 2008, pp. 69–75.
22. Ying-feng, L., S. Zhongke, and Z. Zhirna. Estimation of Probability of Swarming Pedestrians Violation at Signalized Intersections in Developing Cities. *Journal of Southwest Jiaotong University*, Vol. 17, No. 1, 2009, pp. 80–85.
23. Hashimoto, Y., Y. Gu, L. T. Hsu, M. Iryo-Asano, and S. Kamijo. A Probabilistic Model of Pedestrian Crossing Behavior at Signalized Intersections for Connected Vehicles. *Transportation Research Part C: Emerging Technologies*, Vol. 71, 2016, pp. 164–181.

The Standing Committee on Transportation in the Developing Countries (ABE90) peer-reviewed this paper (18-03775).

Article

TRR
JOURNAL OF THE TRANSPORTATION RESEARCH BOARD

Performance Comparison of Bus Travel Time Prediction Models across Indian Cities

Transportation Research Record
2018, Vol. 2672(31) 87–98

DOI: 10.1177/0361198118770175
journals.sagepub.com/home/trr
SAGE

Jairam R[1], B. Anil Kumar[2], Shriniwas S. Arkatkar[1], and Lelitha Vanajakshi[2]

Abstract

Road traffic congestion has become a global worry in recent years. In many countries congestion is a major factor, causing noticeable loss to both economy and time. The rapid increase in vehicle ownership accompanied by slow growth of infrastructure has resulted in space constraints in almost all major cities in India. To mitigate this issue, authorities have shifted to more sustainable management solutions like Intelligent Transport System (ITS). Advanced Public Transportation System (APTS) is an important area in ITS which could considerably offset the growing ownership of private vehicles as public transport holds a noticeable mode share in several major cities in India. Getting access to real-time information about public transport would certainly attract more users. In this regard, this work aims at developing a reliable structure for predicting arrival/travel time of various public transport systems under heterogeneous traffic conditions existing in India. The data used for the study is collected from three cities—Surat, Mysore, and Chennai. The data is analyzed across space and time to extract patterns which are further utilized in prediction models. The models examined in this paper are *k*-NN classifier, Kalman Filter and Auto-Regressive Integrated Moving Average (ARIMA) techniques. The performance of each model is evaluated and compared to understand which methods are suitable for different cities with varying characteristics.

Mobility is the freedom and ability to travel as fast as possible from a particular origin to a desired destination. Over the last two decades, rapid urbanization and economic development has led to a noticeable increase in motorized vehicles. India, like many other countries, is suffering from this unwelcome change leading to slow moving traffic, hour-long queues and longer commutation times. This has adversely affected the mobility of commuters in the country. As more and more vehicles get added to the existing road network, traffic congestion has become a decisive factor for indexing the quality of life in a given city. In this regard, there is a need for better traffic operation of public transport, making it a prime choice for daily commuters. Public transport is important for ensuring sustainable development in the country (*1*). To attract more people toward public transport, it should provide quality service to passengers. Providing real-time bus arrival information to passengers can make bus transport user-friendly and enhance its competitiveness with other modes of transport (*2, 3*). With a schedule of predicted arrival times at each bus stop available via display boards or as mobile or web application, people can make timely plans for future trips. This needs a way to accurately predict the travel time of buses. This study is focused on this prediction application, which could attract more passengers to use public transport, which in turn can lead to less traffic congestion.

The revolution of technology has brought immense change in the quality and frequency of useful data. In the transportation engineering domain, this has resulted in several advancements, mainly under the umbrella of Intelligent Transport System (ITS). One example is city transport buses being deployed with Global Positioning System (GPS) devices, which generate real-time vehicle location data every 1–10 seconds. This massive amount of vehicle trajectory data can be used for real-time travel time prediction and fleet management. Regardless of the amount of data collected, the prediction accuracy depends significantly on the quality and significance of the input. Time-tagged location data, usually represented in the form of trajectories, bring a great potential for real-time prediction of the vehicle travel times. The travel condition of a bus may easily be affected by various internal and external factors, including accidents, weather, road construction, government policies and weather. Also, for GPS fitted vehicles, errors often exist in positional data acquisition because of interference by urban canopies and other reasons.

[1]Department of Civil Engineering, Sardar Vallabhbhai National Institute of Technology, Surat, Gujarat India
[2]Department of Civil Engineering, Indian Institute of Technology Madras, Chennai, India

Corresponding Author:
Address correspondence to Shriniwas S. Arkatkar: sarkatkar@ced.svnit.ac.in

In this study, the challenges mentioned above are considered, and three prediction models are proposed incorporating the vehicle-based GPS data collected from buses from three different cities in India—Surat, Mysore, and Chennai. The data collected for the study includes data from bus-only lane network of Bus Rapid Transit System (BRTS) operated by SITILINK Ltd., mixed traffic bus lane from Mysore operated by Karnataka State Road Transport Corporation (KSRTC), and from Metropolitan Transport Corporation (MTC) in Chennai.

In general, approaches on travel time estimation and prediction using vehicle-based data can be broadly classified into two main classes—data-driven methods and model-based methods. Some of the most important studies which were carried out over the past few decades in these areas are reviewed here.

The various reported techniques for bus travel time prediction include prediction using average speed techniques (*4*), step-wise linear regression techniques (*5*), time-varying coefficient (TVC) linear regression model techniques (*6*), time series analysis techniques (*7–8*) and filtering techniques (*3, 9*). With the help of faster computers and approachable coding languages, researchers have widely explored machine learning techniques in real-time arrival/travel time prediction. Among several artificial intelligence techniques, the most significant methods include Artificial Neural Network (ANN) (*10*), Support Vector Regression (SVR) (*11*), and *k*-Nearest Neighbor (*k*-NN) classifier (*12*). In most of these machine learning techniques, the model learns the travel patterns existing in the field by extracting critical parameters such as travel time, segment speed, traffic volume, and so forth from historical data.

Even though these machine learning techniques were cultured enough to understand the traffic behavior, it requires a significant amount of training time to prune the models. Thus, the application of these models to real-time prediction scenarios would be difficult. Recently, researchers have explored the concept of similarities in historical trajectories and present trajectory of a bus, for predicting travel time on a real-time basis (*13*). The studies showed that historical trajectory-based travel time prediction is highly accurate when a large amount of historical data is available. The historical data inherits the stochastic features of traffic for each hour of the day, thus enabling the model to accommodate for most possible changes in the traffic conditions.

Due to the inconsistent nature of traffic, the travel time will not be the same over different time periods, for the same network. Several studies have been carried out to describe this travel time variability with the help of several distribution functions (*14–17*).

From the literature review, most of the studies are concentrated locally on the application of each method on a particular system of public transport. The results observed are constrained to a particular set of data. Thus, the comparison of each method would not reflect its true potential. This study compares the performance of three prediction techniques using the same data from three different site conditions. Such a uniform comparison of various methods across various cities has not previously been reported for bus travel time prediction, especially under Indian traffic conditions.

Study Area and Data

The data used in this study were collected using GPS devices fitted inside buses from the three cities. Each bus is equipped with a GPS device that records the location status of the bus along with its movement. These data are transferred to a central server every 10 seconds. Three different routes were considered for this study:

1. Bus only corridor of 9 Km in Surat including 16 stops between Udhna Dharwaja and Sachin GIDC
2. Mixed lane traffic of 7 Km in Mysore with 12 stops between City Bus Stand and J P Nagar
3. A 27 Km long route with mixed traffic and 15 stops between the source, Saidapet Bus Stand, and destination, Kelambakam Bus Stand

The routes mentioned above are shown in Figure 1. These three routes characterize most possible combinations of heterogeneous traffic and land use that would comprise a public transport corridor. Each city being different in its own demography and culture, the data collected from these would help to test the robustness of the models used for prediction. The variation in the results from each model in these routes would highlight the advantage of each model against the other. This could act as guidelines to authorities in selecting suitable prediction models when applying them in the field.

With the help of the positioning technology, the GPS device records the current status of the bus in terms of longitude, latitude and time stamp. Each bus has its own identification number. The routes traversed by the bus are recorded with a unique Trip ID. It also contains information regarding multiple trips made by that bus in that day. Each trip was named in the format *"<date>.<direction><trip start time>"*, where date corresponds to the date of data, start time is the time at which the trip is initiated, and direction implies whether the trip is onward or return. For identifying the start and end of the trip, longitude and latitude of origin and destination stops are used. When the bus reaches the destination point, velocity recorded by the device will turn to zero for a specific period, that is, for the duration in which the bus is stationary at the depot. This helps in terminating the trip while the same is cross-checked with the latitude and longitude.

Once the trip-wise data is identified, the distance between each GPS data pair is calculated using Haversine formula (*18*). The difference between consecutive time stamps gives the time difference (Δt) between each location data sent from the GPS device. Since segment-wise travel time is more

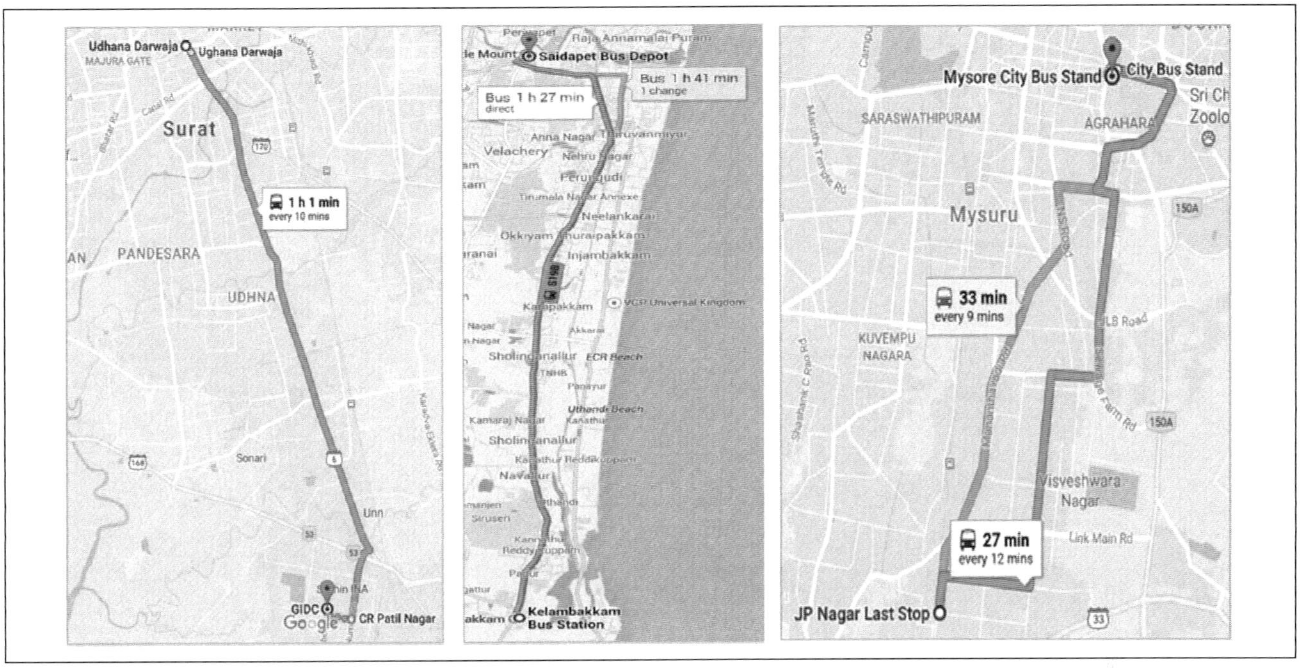

Figure 1. Study routes: (*a*) Surat (left), (*b*) Chennai (middle), and (*c*) Mysore (right).

practically useful to passengers than at a route level, the distances calculated between GPS time stamps are further converted to segment-wise travel time data. For a particular route, these segmental travel times were stored in a grid layout where each column represented a trip and each row represented a segment. Thus, a column consisted of the travel times on a particular trip for all the segments and a row consisted of the travel times on all the trips of the route for a particular segment. For this study, the routes were discretized into smaller segments of length 300 meters each. These segment-wise travel time data form the inputs to all the prediction methods which are detailed in the next section.

Travel Time Pattern Analysis

Traffic in any roadway is expected to follow certain temporal and spatial patterns. Temporal patterns can include hourly pattern (peak and off-peak), daily pattern (weekday vs. weekend), weekly pattern (same days of the week having similar pattern) and so forth, whereas spatial patterns may be specific to certain sections or adjacent sections. The temporal variation in travel may be because of the change in land use during different time periods. Also, for different cities, the working hours might be different based on the population and industrial background of the city. To understand these influences in bus travel time, the data collected from three cities are studied to find the travel time variation over time. Identification of these patterns in the data will help in identifying the best inputs to be used for the prediction application.

To start with the temporal analysis, it is important to identify peak and off-peak hours of a day. For this, the travel time was plotted against each service hour of the day for each study area using box plots. Figure 2 illustrates the travel time variation over different hours of the day for Surat, Mysore and Chennai city. The peak hours from these plots were recorded as 9:00–11:00 a.m. and 6:00–8:00 p.m. as peak hours for Surat BRTS, 7:00–9:00 a.m. and 5:00–8:00 p.m. for Mysore City and 8:00–11:00 a.m. and 5:00–8:00 p.m. for Chennai City. The box plots depict the variation in the hourly travel for each route. The data from Chennai city was observed to have high variability compared with Surat BRTS and Mysore city data. The variation over the day for each city's datasets shows the challenges in predicting the travel behavior as it varies for different hours and for different cities. To understand the travel time variability, the present study explores statistical distributions.

Travel time patterns are often represented as trajectories, which give better understanding of travel time variation over space and time. Therefore, in this study, trajectory of each trip completed by the bus is plotted in a space–time field to understand the travel patterns. This helps to identify similar patterns for each case, using which temporal relations can be tested. For comparing the bus trajectories for temporal patterns, the hourly patterns (peak and off-peak hours) were plotted. These trajectory plots for peak and off-peak are shown in Figure 3 in the order Surat, Mysore and Chennai. The X-axis of the plots corresponds to each segment of the route, while Y-axis contains travel time in seconds. The plots show clear difference between peak and off-peak hour trips, with the peak hour trips occupying upper half of the spectrum. From Figure 3, it can be observed that the trips that happened in the peak hours are almost together at upper part

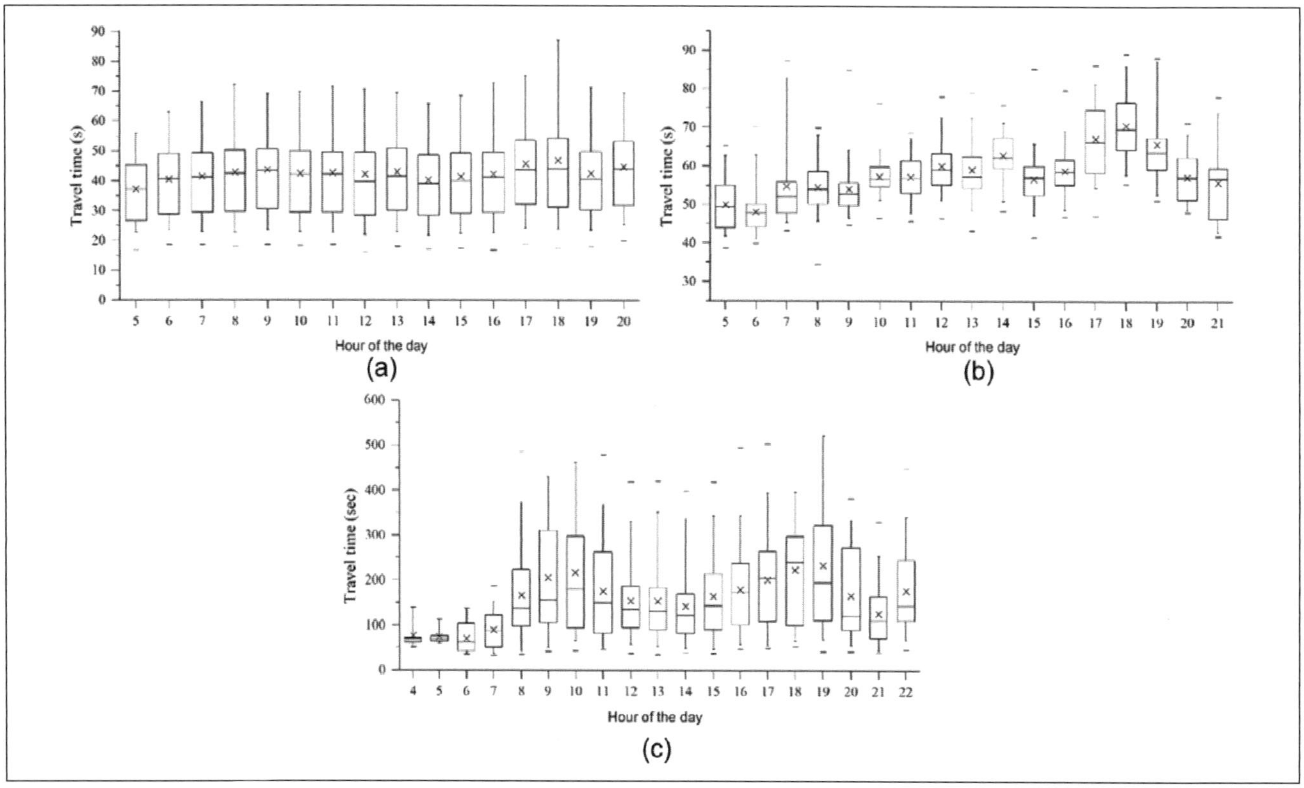

Figure 2. Hourly box plot for (*a*) Surat, (*b*) Mysore, and (*c*) Chennai.

of the figure (i.e., the trips are taking more time to cover the route) and the trips that happened in the off-peak hours are at lower part of the figure (i.e., these trips are taking relatively less time than peak hour trips to cover the route), showing a clear difference in variation in travel time. To confirm this, an alternative statistical analysis was also carried out by fitting the distributions to travel time data. As per the literature, travel time can be explained using three statistical distributions—Burr, Log normal and Generalized Extreme Value (GEV) (*14–17*). In this regard, distribution fitting has been carried out at three levels—peak, off-peak hour, and day of the week. GEV distribution is from a family of normal distribution and it consists of three parameters, namely shape parameter (k), location parameter (s) and size parameter (μ). The parameters of GEV, particularly shape parameter, may be employed in identifying the variability of the travel time. Easy Fit Professional software (*19*) was used for distribution fitting in this study. The software runs the data through 60 probability distributions and provides the list of curves along with the best fitting curve using the Kolmogorov-Smirnov (KS) Test. Figure 4*a–c* shows the GEV plots for the peak and off-peak hour data for each city in the study.

Along with the hourly analysis, an additional analysis was carried out to identify the daily patterns in travel time data. The plots in Figure 4*d–f* show the daily pattern in travel time data. As Surat data contain BRTS data which is bus-only lane-based travel data, the variation is low, with the GEV closely spaced between each other. In Mysore, the travel time data is mixed traffic with lower vehicular population in the city, which reflects in the GEV PDF with noticeable difference in the peak of each curve. Moreover, the base of GEV which shows the variance in data for each curve is not extensively flat which shows that the data does not vary much within these hours. In the case of Chennai, the mixed traffic data and larger vehicular population in the city results in noticeable deviation between the peaks of each GEV PDF in addition to larger base.

From the preliminary analysis, the variation in travel time for each period is identified. The travel pattern varies for peak and off-peak hours and through the week it varies for each day. The plots help in identifying the dissimilarity in trajectories for each level of analysis, which is required in optimizing the historical data that is entered as input to each prediction model for easier and faster computation.

Prediction Models

k-NN Classifier

In this study, the historical trajectories are first classified based on the similarity in pattern with the input data. This was carried out using a non-parametric learning algorithm called *k*-NN classifier and the advantage with this method is that it does not make any assumptions on the underlying

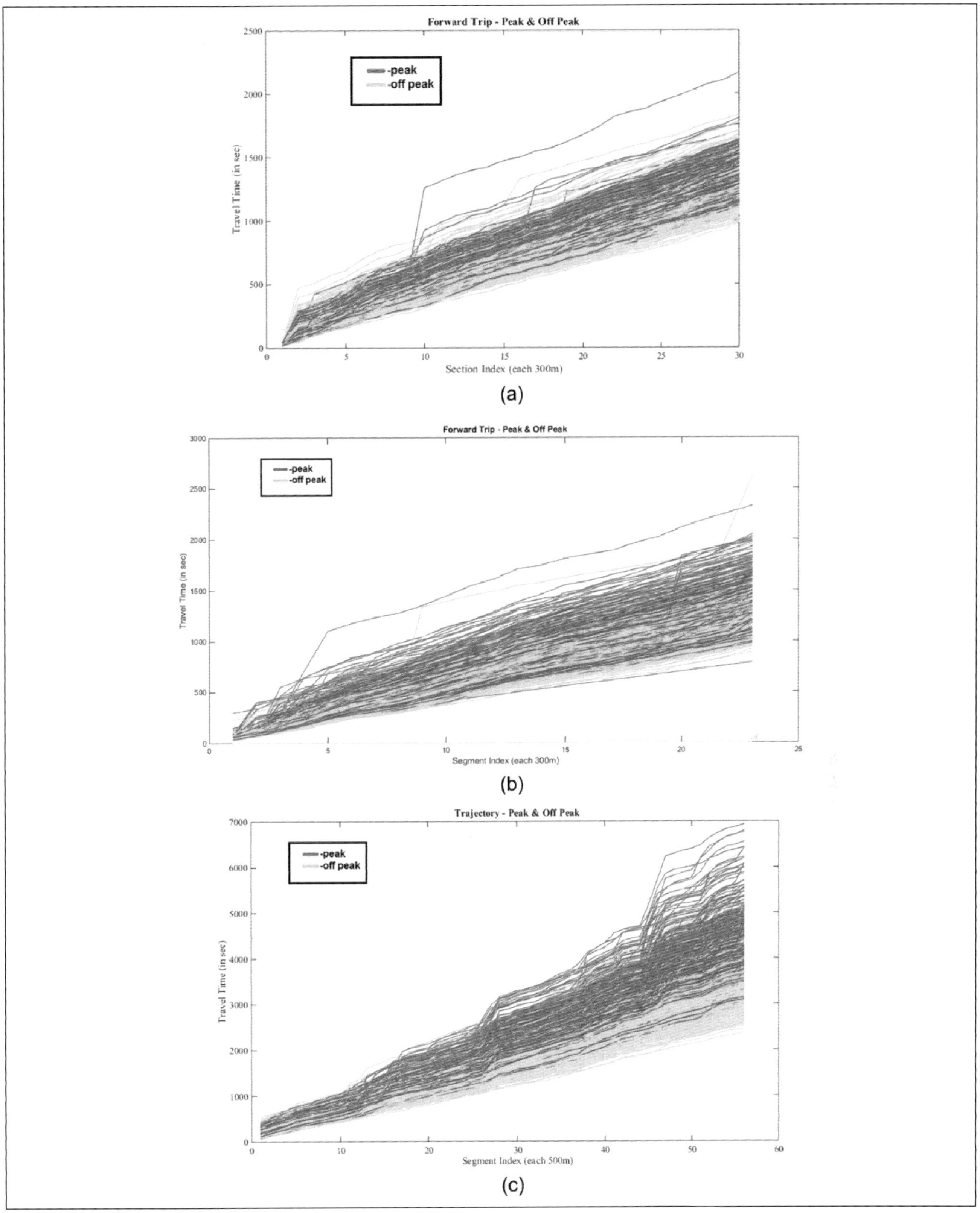

Figure 3. Hourly trajectory plot: (*a*) Surat, (*b*) Mysore, and (*c*) Chennai.

data, nor does it use any explicit training data to do any generalization, or it is very minimal. *k*-NN uses Euclidean distance between the input and historical data as a parameter for classifying the dataset, where *k* represents the number of neighbors. The algorithm finds the nearest neighbors based on Euclidean distance of each historical data. The class

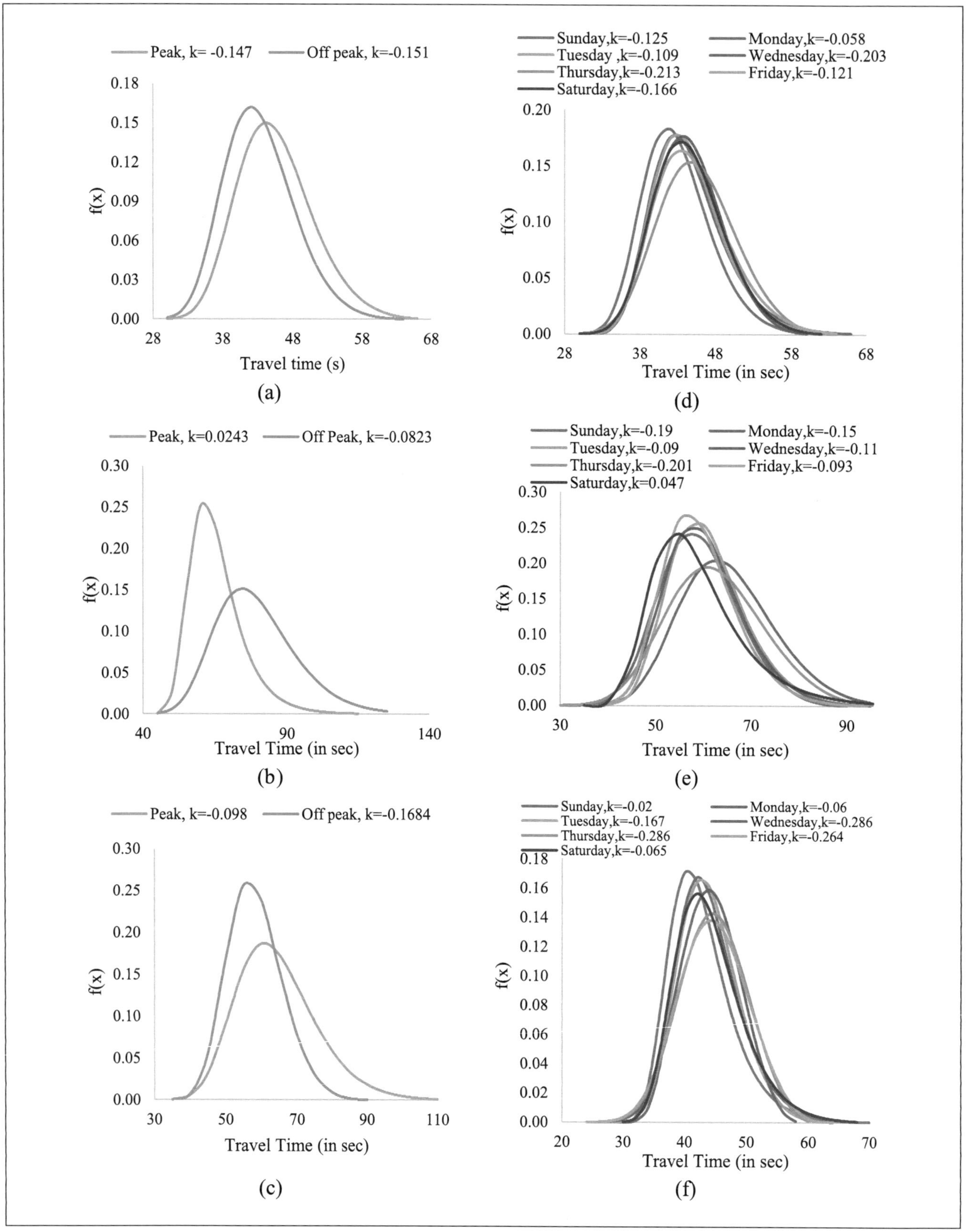

Figure 4. GEV PDF comparison for hour of the day in (*a*) Surat, (*b*) Mysore, and (*c*) Chennai, and day of the week in (*d*) Surat, (*e*) Mysore, and (*f*) Chennai.

Table 1. Sample k-NN Classifier Output for $n = 4$ and $k = 10$

Nearest neighbor	n-4	n-3	n-2	n-1	n	Euclidean distance
1	45.26	63.69	33.37	70.86	69.90	26.53
2	30.46	60.47	36.17	53.07	56.90	27.77
3	44.85	56.64	29.07	59.91	61.00	28.69
4	38.29	90.31	29.28	53.08	82.46	29.03
5	35.78	73.64	44.85	56.64	64.60	29.07
6	28.69	70.66	38.29	90.31	87.02	29.28
7	32.36	57.61	33.27	70.00	66.32	29.78
8	35.84	50.35	34.47	55.11	51.68	30.46
9	26.53	46.53	33.06	55.38	52.46	30.51
10	29.28	53.08	29.03	97.64	87.73	30.80

which the output is closest to is determined by the majority voting on the class labels of its k-NNs (*20*).

The proposed prediction method being a real-time framework, the input to the classifier algorithm needs to be the patterns from the current trip. To identify such patterns, the travel time of previous 'n' trips were taken as input. For example, if $n = 4$, for predicting the pth trip, the input to the algorithm will be the travel time of the (p-1)th, (p-2)th, (p-3)th and (p-4)th trips. The 'similar' trips from the historical data are then identified based on the Euclidean distance between the current trip travel time and the historical data, thus forming the cluster. Thus, there will be only one cluster formed containing 'k' historical trajectories that are closely related or 'similar' to the given input. The average value of predicted travel times from this cluster is taken as the predicted value for the current input. The algorithm updates for each section with the new inputs of travel time for the same section from previous trips. This will accommodate for the changes in traffic in the field, making it change in real time. Sample results of k-NN classifier using $n = 4$ is shown in Table 1. Here, the parameters considered for the k-NN classifier algorithm are $l = 300$ m, $k = 10$ and $n = 4$. In this case, for predicting the nth trip, the previous four trip travel times are taken as input and the results for each combination are sorted with respect to the Euclidean distance, and k nearest neighbors are considered for averaging.

Auto-Regressive Integrated Moving Average (ARIMA)

Box and Jenkins (*21*) had carried out thorough investigation on the analysis of time series (TS) and put forward well-structured classes of models, such as Auto Regressive (AR), Moving Average (MA), Auto Regressive Moving Average (ARMA), Auto Regressive Integrated Moving Average (ARIMA), Seasonal Auto Regressive Integrated Moving Average (SARIMA), and so forth for modeling TS data. In this study, the travel time is represented as segment-wise travel time. Since segment-wise travel time data are not TS as they do not happen at uniform intervals, an approach was adopted to take the total travel time for individual trips. Since each trip starts with respect to certain time of day, the total travel time can be considered as TS data.

An ARIMA model consists of four stages—the model identification, model estimation, diagnostic checking and forecasting. To start with, the data is checked for stationarity as the data set used in TS analysis is assumed to be stationary. If it is not stationary, stationarity is achieved by differencing the data. Figure 5, *a* and *b* illustrate a sample raw data and the same after stationarity. Once the degree of stationarity is defined, the order of AR and MA process is identified using autocorrelation function (ACF) and partial autocorrelation function (PACF). Figure 5, *c* and *d* show the PACF and ACF plot for a sample data. Correlation values range from −1 to +1. A value of +1 indicates that the two variables move together perfectly; a value of −1 indicates that they move in opposite directions. When building a TS model, it is important to include lag values that have large, positive autocorrelation values. Sometimes it is also useful to include those that have large negative autocorrelations. Thus, from Figure 5, *c* and *d*, the order of ACF is 1 while order of PACF can be between 1 and 4.

Once the model parameters are prepared, the models are run with route-level travel time as input, which is equal to the sum of travel time of all segments of the route. The forecasting accuracy in ARIMA is better for short term prediction (*4*). Thus, forecasting is carried out for each model with few steps at a time. The predicted values then feed back into the raw data as historical data. Since this study concentrates on predicting travel time at segment level, the predicted route level travel time are then decomposed to segment data. For decomposing, certain segment-specific coefficients are used, which were established using historical database. Travel time on each segment for a trip is divided by the corresponding total travel time, thus obtaining the segment-wise coefficients for the trip. These coefficients are formulated for each hour of the day for every route considered in this study. These segment-specific coefficients take into account the route travel time variations for each hour of the day, thus accommodating the dynamic nature of traffic which varies temporally and spatially.

Kalman Filtering Technique

The third prediction approach explored in this study is using Kalman Filtering Technique (KFT). KFT uses the output from k-NN, that is, similar trajectories as input to predict the bus travel time. The implementation of KFT requires information regarding the system's dynamics, statistical information of the system disturbances and measurement errors. The evolution of travel time between various time intervals in a given subsection is assumed to be

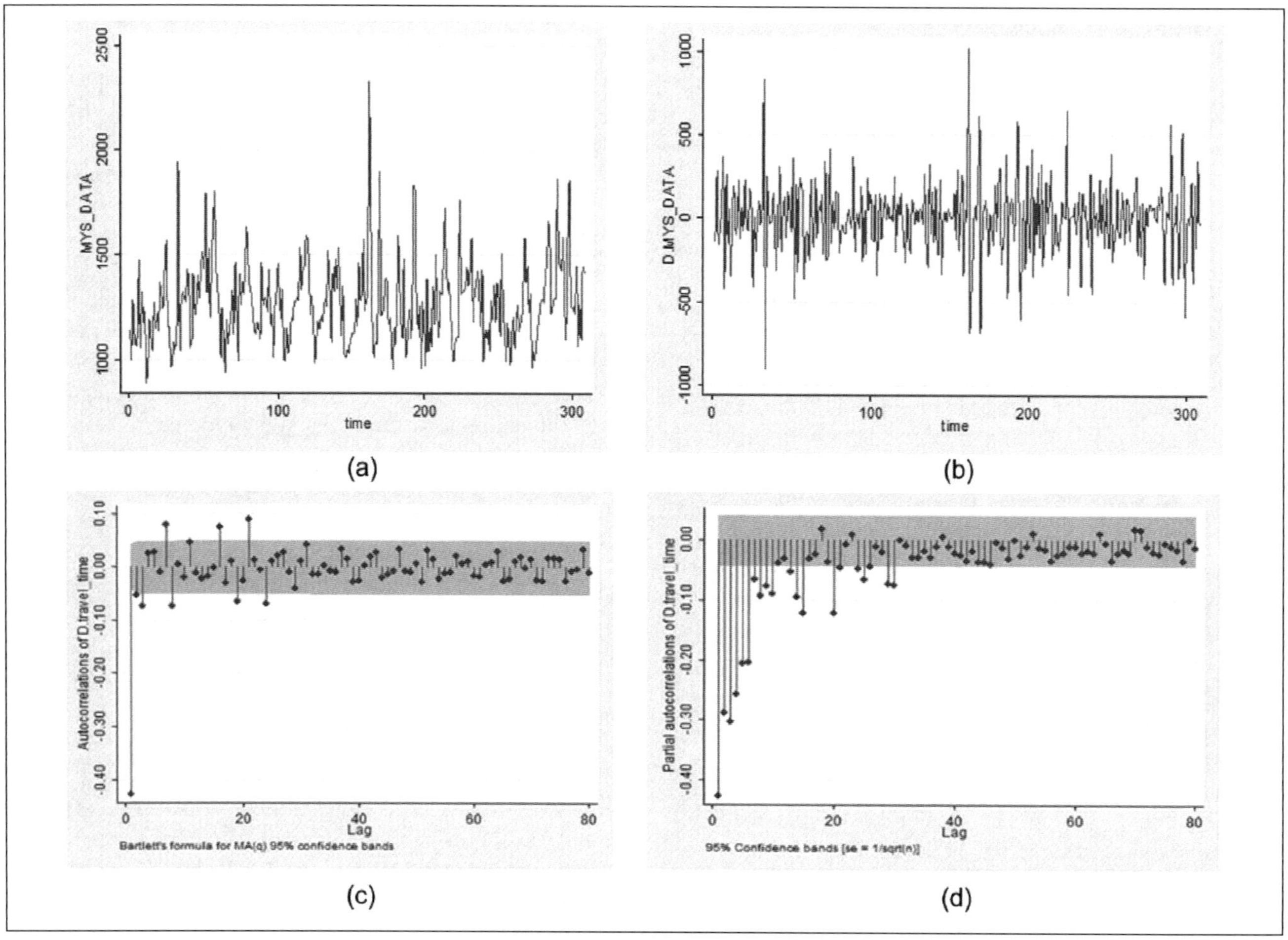

Figure 5. ARIMA Plots: (*a*) raw data, (*b*) stationary data, (*c*) ACF, and (*d*) PACF.

$$x(t+1) = a(t)x(t) + w(t), \tag{1}$$

where $a(t)$ is a parameter which relates the time taken to travel in a given subsection, $x(t)$ is the travel time taken for covering the given subsection at time t and $w(t)$ is the associated process disturbance. The measurement process was assumed to be governed by

$$z(t) = x(t) + v(t), \tag{2}$$

where $z(t)$ is the measured travel time in a given subsection at time t and $v(t)$ is the measurement noise. The algorithm requires two sets of data (S1 and S2) to implement the above scheme. Out of these two data sets, one set of data (S1) was used in the time update equations to calculate the parameter $a(t)$ and the other set of data (S2) was to be used in the measurement update equations to generate the *aposteriori* travel time estimate. So, the results obtained from *k*-NN classifying algorithm were arranged as two sets of data in order of preference from lower to higher (higher preference being the data that is nearer to the input pattern). The algorithm will be repeated for all N subsections. The objective here is to predict the travel time of the Test Vehicle (TV) by identifying significant travel time trajectories in a given subsection.

Results

In this section, the performance of each model is evaluated and compared for the three cities. The prediction window is one week for each city which gives clear understanding of how the models perform for different days of the week. For a particular bus in an ongoing trip, the predicted segmented travel time (TT_{pred}) using prediction models are stored in the database. The observed segmented travel time (TT_{obs}) are collected from the historical data. For evaluating the accuracy of prediction, Mean Absolute Percentage Error (MAPE) and Mean Absolute Error (MAE) were used as measures and are calculated as in Equations 3 and 4. The lower the errors, the more accurate are the predicted travel times and hence, the better the method.

$$\text{MAPE} = \frac{1}{n}\sum_{i=1}^{n} \left| \frac{(TT_{obs} - TT_{pred})}{TT_{obs}} \right| * 100, \tag{3}$$

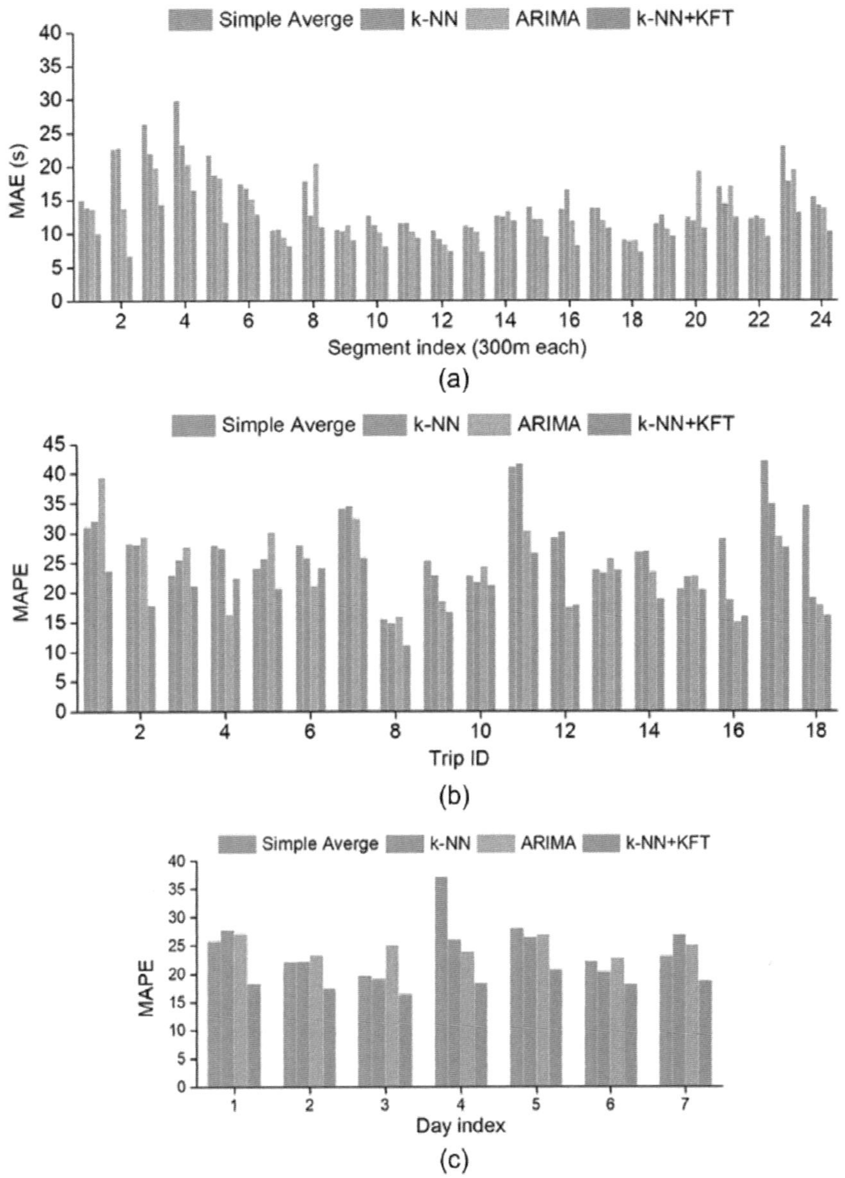

Figure 6. Performance comparison between models for Mysore city: (*a*) segment-wise, (*b*) trip-wise, and (*c*) day-wise.

$$\mathrm{MAE} = \frac{1}{n}\sum_{i=1}^{n} |\,\mathrm{TT_{obs}} - \mathrm{TT_{pred}}\,|, \tag{4}$$

Then, the performance evaluation was done at three levels—segment-wise, trip-wise, and day-wise. Performance of these methods was also compared with a base method, namely Simple Averaging Technique (SAT). This approach is carried out section-wise, where the average of the travel time of previous '*n*' trips was taken as the predicted travel time for that section.

The performance comparison for Mysore city is shown in Figure 6*a–c*. In Figure 6*a*, segment-wise average of 100 trips were taken and the *k*-NN + KFT performs relatively better than the other models with an average error of 10.21 seconds. The next most accurate model was found to be ARIMA with an average error of 13.79 seconds. The error for *k*-NN + KFT varied between 6.6 seconds and 16.48 seconds while for ARIMA it ranged between 8.4 seconds and 20.4 seconds. Figure 6*b* shows the accuracy of prediction for all trips in a sample day to understand how the models perform over a working day. Figure 6*c* shows a comprehensive summary of how the models perform over each day of the week. Here also, *k*-NN + KFT performs better than other models. It can also be seen that all the proposed methods performed better than the baseline SAT.

Figure 7*a–c*, show a similar comparison for Surat city BRTS corridor. Unlike in the previous case, the segment-wise comparison in Figure 7*a* shows that all proposed

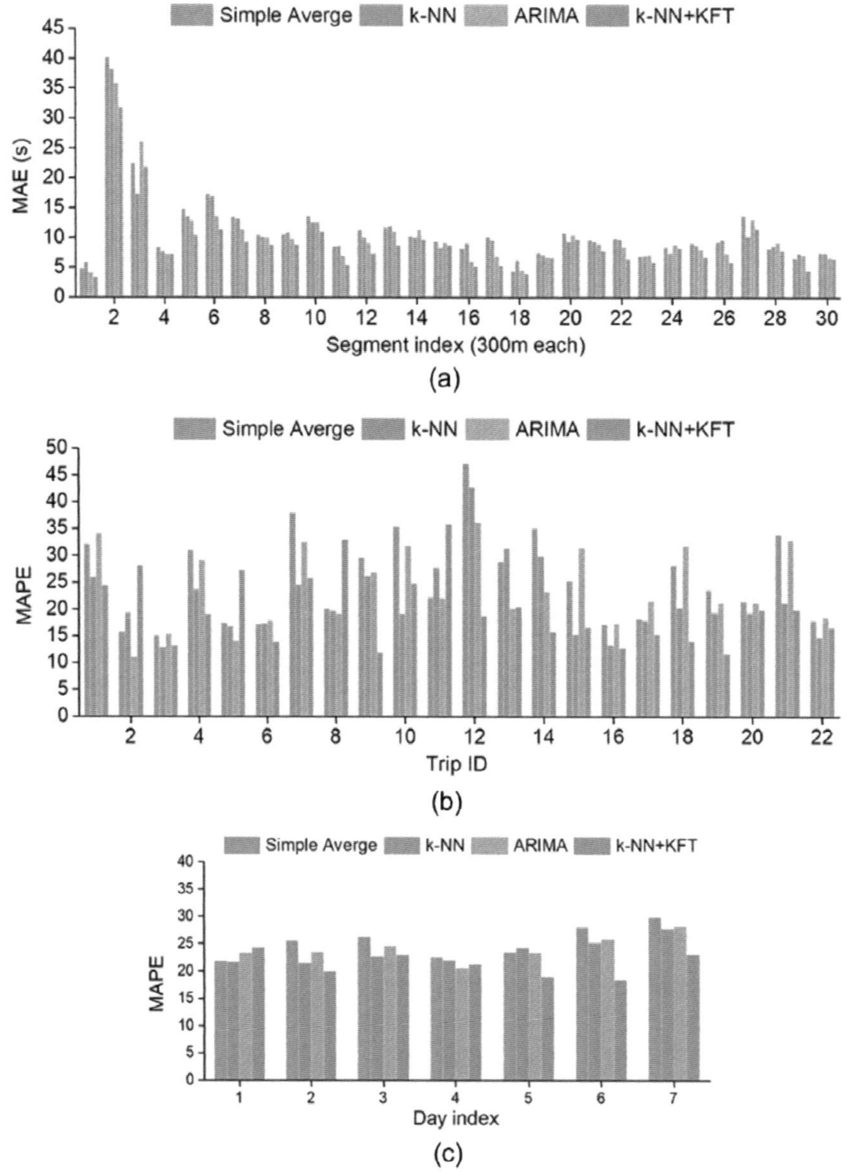

Figure 7. Performance comparison between models for Surat city: (*a*) segment-wise, (*b*) trip-wise, and (*c*) day-wise.

models performed comparably, and only slight improvements were observed when compared with the SAT. Segments 2 and 3 are the two high variance sections in this route, due to a major junction in the route where cross moving traffic interferes with the movement of BRTS. The performance of proposed models can be found to be much better in these sections compared with SAT. Although the segment-wise comparison shows only slight difference between the proposed models, Figure 7*c* gives better clarity on the performance through the prediction window. Except Sunday where the traffic would be generally low, the proposed models performed well for other days of the week. Thus, it can be concluded that for bus-only lane system, the sophisticated models might not be necessary. Overall, *k*-NN+KFT has the lowest error of 18.3% on Friday and the maximum error is on Sunday, with 24.21%.

In Chennai city, the performance comparison was carried out for 312 trips. The bus travel time data for Chennai were highly stochastic due to larger vehicle population observed in the metropolitan city. The heterogeneity in traffic composition was also observed to be highest in Chennai compared with other cities considered in this study. Figure 8*a*–*c* portrays the performance comparison of each prediction model carried out on Chennai city data.

In Figure 8*a*, out of 93 sections of the route, *k*-NN+KFT dominated majority of the sections with high accuracy ranging between 1.6 seconds and 58.65 seconds with a total average error of 9.13 seconds. It is followed by *k*-NN

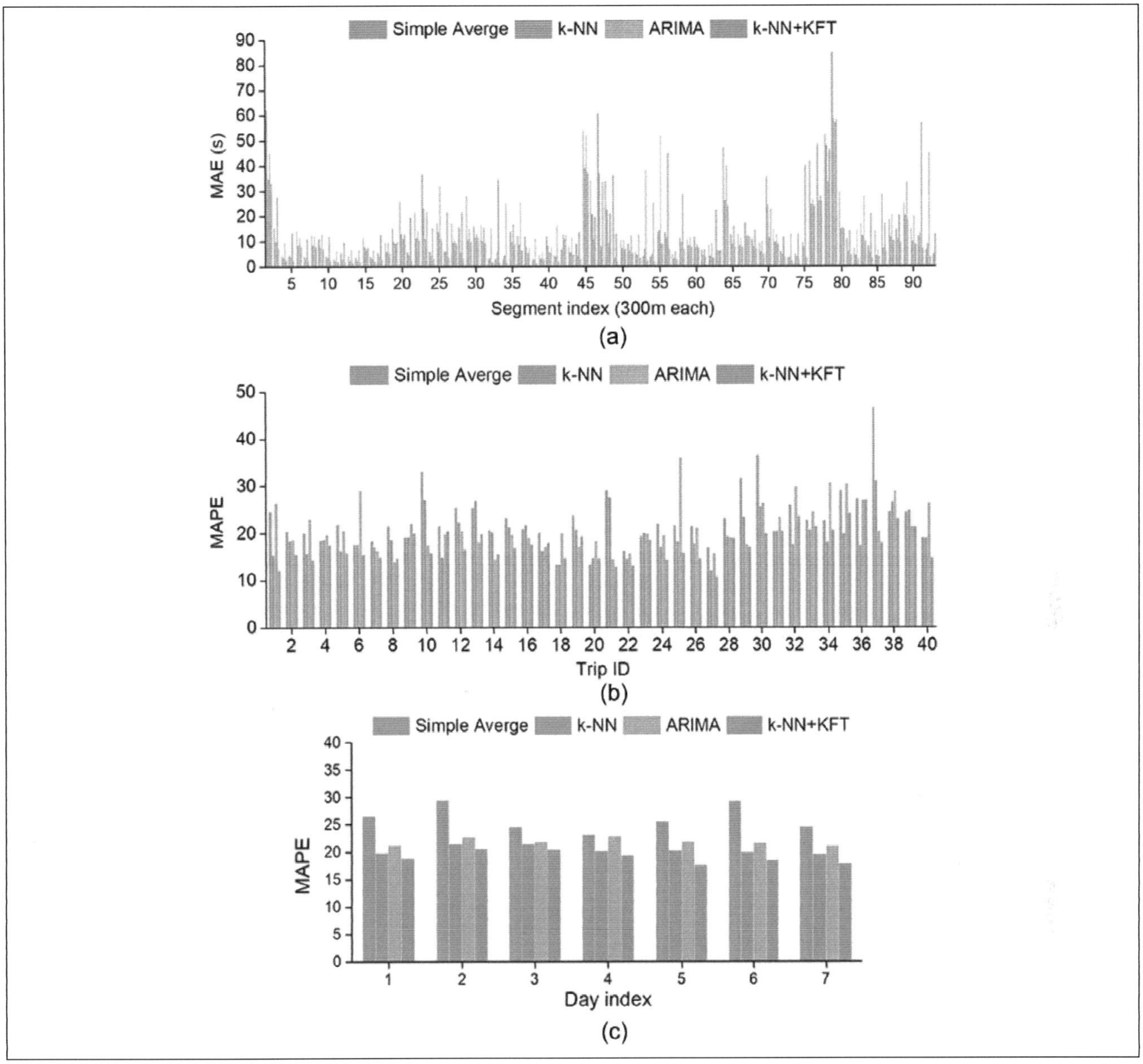

Figure 8. Performance comparison between models for Chennai city: (*a*) segment-wise, (*b*) trip-wise, and (*c*) day-wise.

classifier algorithm with an overall average error of 10.68 seconds, while the segment-wise error varied between 2.47 seconds and 58.85 seconds. Similar to the high variance sections of BRTS corridor, the *k*-NN+KFT model performs well in the sections 76–79, which were the high variance section in Chennai's 19B route. Figure 8*b* shows the comparison of prediction results for a sample day where *k*-NN+KFT performs better compared with other models. The day-wise comparison of prediction results can be seen in Figure 8*c*, which also shows that *k*-NN+KFT performs the best followed by *k*-NN classifier. The error for *k*-NN+KFT varies between 17.5% on Monday and 20.65% on Friday. Thus, *k*-NN+KFT has less error both spatially and temporally compared with other models proposed in the study.

Conclusion

In this paper, the problem of real-time prediction of bus arrival/travel time using historical travel time data is studied. For implementing a prediction framework, the authorities should understand the limitations and advantages of the models and type of inputs required for the purpose. This study concentrates on showing how each model performs under different scenarios, namely bus-only lane and mixed traffic. The study explores three prediction models which used segment-wise bus travel time data as input. To start with, GPS data containing bus travel time information were collected from three cities—Surat, Mysore and Chennai. Surat has bus-only lane while Mysore and Chennai have mixed lane for buses. The data collected from these three cities were analyzed to test the robustness of the proposed prediction methods. First, a pattern analysis of bus travel time data was carried out. Analysis was done for various hours of the day and days of the week. The peak hours were identified and it was observed that peak and off-peak trips have distinctly different characteristics. The travel time data was next investigated for the statistical distributions and GEV was found to be the best fit. To compare the prediction accuracy of the models, MAPE and MAE were used as measures. The predicted values were compared with the observed travel times to understand the level of accuracy. Also, a baseline SAT was used in the comparison to identify the extent to which the models performed. For Surat BRTS data, the comparisons showed that, on the whole, all models performed similarly. Even though *k*-NN+KFT performed better in each case spatially and temporally, the difference in accuracy was small. This might be because BRTS are less influenced by external traffic except in intersections where the bus lane is exposed to mixed traffic. Since the *k*-NN, ARIMA and *k*-NN+KFT require computing power, the authorities can opt for the SAT in this case. However, in the case of mixed lane traffic, kNN+KFT performed better than the other models. Finally, the travel time information using the proposed methods can be expressed in terms of estimated arrival times for each bus and shared with the users through display monitors at bus stops, mobile application, and so forth.

Acknowledgments

The authors acknowledge the Centre of Excellence in Urban Transport at IIT Madras; SITILINK Ltd., Surat; and Mitra ITS wing, KSRTC for the GPS data.

References

1. Shimamoto, H., N. Murayama, A. Fujiwara, and J. Zhang. Evaluation of an Existing Bus Network Using a Transit Network Optimization Model: A Case Study of the Hiroshima City Bus Network. *Transportation*, Vol. 37, No. 5, 2010, pp. 801–823.
2. Meng, Q., and Z. Liu. Trial-and-Error Method for Congestion Pricing Scheme under Sideconstrained Probit-Based Stochastic User Equilibrium Conditions. *Transportation*, Vol. 38, No. 5, 2011, pp. 819–843.
3. Vanajakshi, L., S. C. Subramanian, and R. Sivanandan. Travel Time Prediction under Heterogeneous Traffic Conditions Using Global Positioning System Data from Buses. *IET Intelligent Transportation Systems*, Vol. 3, No. 1, 2009, pp. 1–9.
4. Zhu, T., F. Ma, T. Ma, and C. Li. The Prediction of Bus Arrival Time using Global Positioning System Data and Dynamic Traffic Information. *Proc., Wireless and Mobile Networking Conference*, IEEE, Toulouse, France, 2011.
5. Kwon, J., B. Coifman, and P. Bickel. Day-to-Day Travel-Time Trends and Travel-Time Prediction from Loop-Detector Data. *Transportation Research Board: Journal of the Transportation Research Board*, 2007. 1717: 120–129.
6. Zhang, X., and J. A. Rice. Short-Term Travel Time Prediction. *Transportation Research Part C*, Vol. 11, 2003, pp. 187–210.
7. Al Deek, H. M., M. P. D'Angelo, and M. C. Wang. Travel Time Prediction with Non-Linear Time Series. *Proc., 5th International Conference on Applications of Advanced Technologies in Transportation Engineering*, Newport Beach, Calif., 1998, pp. 317–324.
8. Napiah, M., and I. Kamaruddin. ARIMA Models for Bus Travel Time Prediction. *Journal of the Institution of Engineers Malaysia*, Vol. 71, No. 2, 2010, 49–58.
9. Nanthawichit, C., T. Nakatsuji, and H. Suzuki. Application of Probe Vehicle Data for Real-Time Traffic State Estimation and Short-Term Travel Time Prediction on a Freeway. Presented at 82nd Annual Meeting of the Transportation Research Board,Washington D.C., 2003.
10. Liu, H., K. Zhang, R. He, and J. Li. A Neural Network Model for Travel Time Prediction. *Proc., IEEE International Conference on Intelligent Transportation Systems*, Shanghai, IEEE, 2009, pp. 752–756.
11. Vanajakshi, L., and L. Rilett. Support Vector Machine Technique for the Short Term Prediction of Travel Time. *Proc., IEEE Intelligent Vehicles Symposium*, Istanbul, Turkey, IEEE, 2007.
12. Robinson, S., and J. W. Polak. Modeling Urban Link Travel Time with Inductive Loop Detector Data by Using the k-NN Method. *Transportation Research Record: Journal of the Transportation Research Board*, 2005. 1935: 47–56.
13. Lee, W. C., W. Si, L. J. Chen, and M. C. Chen. HTTP: A New Framework for Bus Travel Time Prediction Based on Historical Trajectories. *Proc., 20th International Conference on Advances in Geographic Information Systems*. California, USA. 2012.
14. Kieu, L. M., A. Bhaskar, and E. Chung. Public Transport Travel-Time Variability Definitions and Monitoring. *ASCE Journal of Transportation Engineering*, Vol. 141, No. 1, 2014, 04014068.
15. Lei, F., F. Wang, G. Lu, and J. Sun. A Travel Time Reliability Model of Urban Expressways with Varying Levels of Service. *Transportation Research Part C. Emerging Technologies*, Vol. 48, 2014, pp. 453–467.
16. Akhilesh. *Study of Travel Time Variability on Bus Routes in Mysore City*. M.Tech thesis report. Sardar Vallabhbhai National Institute of Technology, Gujarat, 2015.
17. Susilawati, M. A. P., S. Taylor, and Somenahalli, 2011. Modelling urban travel time variability with the Burr regression technique. *34th Australasian Transport Research Forum (ATRF)*, Adelaide, South Australia, Australia, 2011.
18. Chamberlain, R. G. *Great Circle Distance between Two Points*. http://www.movabletype.co.uk/scripts/gis-faq-5.1.html. Accessed February 14, 2016.
19. Easyfit 5.6. *Mathwave technologies*, 2004. http://www.mathwave.com
20. Ding, C., and X. He. K-Nearest Neighbor Consistency in Clustering. *Lawrence Berkeley National Laboratory*, Berkeley, Calif., 2004.
21. Box, G. E. P., G. M. Jenkins, and G. C. Reinsel. *Time Series Analysis: Forecasting and Control*, 4th ed. John Wiley & Sons, Inc., Hoboken, N.J., 2013.

The Standing Committee on Transportation in the Developing Countries (ABE90) peer-reviewed this paper (18-04897).

Article

TRR
JOURNAL OF THE TRANSPORTATION RESEARCH BOARD

Transportation Research Record
2018, Vol. 2672(31) 99–109

DOI: 10.1177/0361198118758286
journals.sagepub.com/home/trr
SAGE

How Operators' Legal Status Affects Safety of Intercity Buses in Thailand

Apiwat Ratanawaraha[1] and Saksith Chalermpong[2]

Abstract

Legalization is often a policy solution to perceived problems of inefficient and unsafe operation of informal buses in developing countries. Facing the world's second worst road safety record, the Thai government is set to ban the use of vans for intercity bus services and is considering banning individual operators and requiring them to form a legal entity. A question thus arises whether an operator's legal status significantly affects bus safety. This paper aims to answer that question by analyzing route-level data on accidents and legal status of intercity bus operators. The results of Poisson and negative binomial regression analyses confirm that the use of vans increases accident and fatality risks of intercity bus services. But the evidence does not support the government's claim that requiring operators to register as legal entities would make intercity services safer. For long-distance routes, more bus companies mean greater accident risks. The only situation where more individual operators mean higher safety risks is with fatality rates on long-distance routes. However, fatality rates are actually lower with more individual operators on short-distance routes. This is likely because most individual operators on short-distance routes belong to informal operator groups that adopt internal rules and procedures to govern members' behaviors. The key policy implications are that mandatory legal registration of operators does not automatically enhance bus safety, and that safety measures should focus on operational characteristics and actual practice.

Subscribing to the concept of natural monopoly, the Thai government granted the state enterprise Transport Co. Ltd. (TCL) in 1959 the sole license to operate all intercity bus routes between Bangkok and other provinces. The objectives were to reduce cutthroat competition among independent operators and to provide safer, more efficient, and cost-effective services. However, with growing travel demand as a result of rapid demographic and economic growth, as well as the government's limited financial support, TCL did not have adequate resources to expand its operation to cover all routes. So the government permitted TCL to subcontract its bus operation to private companies and individual operators. Meanwhile, numerous informal passenger vans operate in many routes to serve unmet demand. Over time, as the number of unlicensed vans grew out of control, the government ordered TCL to include the informal vans as individual subcontractors.

Such policy decisions helped increase the number of intercity buses and routes over the years. However, the services have not become safer. On the contrary, bus accidents and the resulting fatalities and injuries have increased in both absolute and relative terms. In the background is Thailand's road safety record, which is the second worst in the world, with 14,059 fatalities in 2012, or 36.2 people per 100,000 (*1*). Although 73% of those killed are motorcyclists, buses and vans account for many of the accidents. In 2015, 394 serious bus and van accidents were reported, resulting in 232 fatalities and 2,711 serious injuries (*2*). Intercity services account for more than 80% of the total accidents, fatalities, and injuries, about 25% of which involve vans.

To enhance bus and van safety, the Thai government has adopted several measures that follow common knowledge and practice regarding determinants of road accidents: namely, human, vehicle and equipment, and environmental factors. In general, they place emphasis on upgrading road and other infrastructure quality, enforcing vehicle standards, and controlling driver behaviors. For bus and van safety in particular, the government bans overloading, while requiring that seatbelts be installed for all seats. The police regularly monitor and fine drivers for reckless behaviors, such as speeding, illegal passing, and drunk driving. All public transport vehicles now have to install a GPS tracking device so that the authority can monitor their speed. Yet accidents still occur frequently, especially during major holidays.

Due to the vans' high accident rates, it has become an established belief among Thai policymakers that the informal nature of van services was *the* root cause of rampant crashes. Bus operators with official and legal entities are perceived

[1]Department of Urban and Regional Planning, Faculty of Architecture, Chulalongkorn University, Bangkok, Thailand
[2]Department of Civil Engineering, Faculty of Engineering, and Transportation Institute, Chulalongkorn University, Bangkok, Thailand

Corresponding Author:
Address correspondence to Saksith Chalermpong: saksith.c@chula.ac.th

to be less likely to cause accidents than independent individual van operators, because they are presumably more organized and have more resources to implement safety measures. Legally organized groups are also thought to offer better compensation and benefits to the drivers, especially when the compensation is not contingent on duration and distance of travel. Based on such thinking, the Thai government has decided to ban the use of vans, and is considering banning individual operators and requiring that they form a group and be registered collectively as a legal entity. The options for legal status being considered include registered companies, limited partnerships, and cooperatives.

Although the safety rationale for requiring legal status seems sound, one could also argue that legal status itself matters less than other factors that help improve operational safety. Individual owner-operators may be safer than bus companies under certain conditions. For instance, some companies may not maintain their vehicles as well as they should, or less diligently than independent owner-operators who drive and take care of their own vehicles. Companies may not pay their employee drivers well enough and possibly less than what independent operators may earn. Companies may also subcontract the whole or part of their operation even further to independent drivers without controlling the service quality and safety. In short, de jure legal status could mean little in terms of safety guarantee. A question thus arises whether an operator's legal status really matters when it comes to safety.

Accordingly, this paper aims to test empirically whether legal status of operators affects safety of intercity bus and van services in Thailand. After reviewing relevant literature in the following section, an overview is provided on how intercity bus and van services are operated in Thailand, and the government's recent policy move to ban the use of vans and to require that individual operators become part of collective legal entities is described. The hypotheses, research methodology, and data are then outlined, and the findings are then presented. The last section concludes with policy implications and suggestions for future research.

Literature Review

Two related strands of literature lay the conceptual groundwork for this paper: determinants of bus safety and safety of informal bus services in developing countries.

Determinants of Bus Safety

Among the vast literature of road safety, the conceptual framework developed by William Haddon in 1970 is widely used for general analysis and policymaking purposes. According to the Haddon Matrix, the main factors contributing to accidents are categorized into three groups (*3*). First, vehicle and equipment factors include braking, lighting, tire quality, and so forth. Second, environment and infrastructure factors include road design and layout, intersection configuration, road curvature, roadway lighting, and so forth. Third, human factors include driver vision, fatigue, knowledge, alcohol and physical impairment, police enforcement, and so forth. Numerous empirical studies have confirmed the relevancy and importance of these factors for bus safety, which could be used as control factors for regression analyses in this study.

Another group of studies explore socioeconomic determinants of bus safety. Compensation levels and methods are found to significantly influence driver behaviors, which in turn affect their involvement in accidents (*4*). Driver characteristics and attitudes also affect bus accident risks. For instance, an analysis of US bus accident databases finds that bus drivers with different characteristics are affected by various risk factors in different ways. Young and elderly drivers with a history of traffic violations are more likely to be involved in accidents in wet road surface conditions than are other groups of drivers (*5*). A study in China using a driver behavior questionnaire finds that among different types of professional drivers, bus drivers are most likely to drive in risky manner and male drivers tended to engage in risky driving behaviors (*6*). Another study from China finds that drivers' tendencies of violations and attitude toward violations contribute to increased accident risks (*7*).

The business structures of bus operators could also affect accident risks. One study finds that owner-operators are more prone to driver and vehicle violations, but have better crash records than employee drivers (*8*). Meanwhile, unionization improves driver and safety performance, which could be attributed to job security and compensation methods (*9*). By examining business structure and drivers' employment status, these studies implicitly test whether operators' legal status is a determinant of bus safety. These studies, however, use data at the individual operator level, so they cannot capture group-level and route-level characteristics.

Some studies on bus safety focus specifically on intercity buses, as they tend to have worse safety records than intracity buses. A study from Malaysia finds that road environment conditions contribute to intercity bus safety risks in a certain region more than others, because of the expansion of the expressway network that can safely accommodate intercity buses. It also finds that different companies show mixed safety performance in different risk domains. The results indicate that bus safety problems cannot be solved by a single road safety intervention, but require targeted, company-specific safety programs (*10*). Other studies have shown that nighttime driving is more dangerous because of driver fatigue and visibility restriction (*11*). Because intercity bus services tend to be of long distance and often during nighttime, such risks are expected to be greater.

Legalization of Informal Transport

Several articles have discussed safety issues of informal buses in developing countries, citing overloading, on-road competition, violation of traffic rules, and poorly maintained vehicles as the causes of accidents and their severity (*12*). These studies usually describe the risk factors without analyzing systematically and showing empirically the causal connections. Data availability could be a reason for limited empirical research on this matter. Among a few studies that examine accident causes in detail, a study in South Africa analyzes a fatal accident involving a minibus (van) taxi, using police accidents reports, site examination, and lab tests (*13*). The study focuses on vehicle and environmental factors that may have caused the accident and fatality. In addition to vehicle problems with brakes and tires, the study finds overloading as a key risk factor. The study also alludes to profitability as an underlying cause of overloading and drivers' dangerous behaviors, but it does not test the hypothesis empirically.

Many risk factors in the informal bus industry in developing countries are attributed to its hyper-competitive nature (*12*). Operators often cut corners by using substandard vehicles, while on-road competition leads to dangerous driving behaviors. In addition, driver compensation is usually based on distance and work hours, and mostly without labor protection and insurance. So drivers work long hours and often at night, which further induces fatigue and greater safety risks. Informality and hyper-competitiveness are thus jointly considered to be the fundamental cause of bus safety risks in developing countries.

As a result, formalization is usually prescribed as the standard policy solution to informal bus problems. One key component of the formalization policy is to require that independent individual operators form a collective legal entity, such as a company. The South African government, for instance, has implemented several policy measures to formalize informal taxi/van services to address safety and other issues. As part of the Taxi Recapitalization Program, taxi operators in Johannesburg were incentivized to scrap their old vehicles in exchange for the allowance that would be invested as working capital in a new bus company (*14*).

Meanwhile, changes in competition models could greatly affect safety levels of bus services. The bus reform in Bogota, Colombia in the early 2000s led to a dramatic decrease in accidents. One key element of the reform was to change the way operators were compensated. Instead of receiving a bonus for the number of passengers transported, operators are paid according to traveled kilometers and service quality. This reduces the need to race on the road for passengers, thus reducing accident risks (*15*). By contrast, in places where regulatory institutions are weak, increased competition through deregulation could worsen accident rates. Delhi's bus liberalization increased the number of private buses from 1,100 in 1992 to 3,800 in 1995, but the number of accidents also rose from 134 to 1,072 during the same period. The number of citations for reckless driving also jumped during that time (*16*). Similarly, bus liberalization in Santiago de Chile in the 1990s increased not only the number of bus fleet but the number of accidents. In this case, even after liberalization, operators' revenue was still based on fare collection, so operators had strong incentives to race on the road. The important lesson to learn from these cases is that the solution lies in the change in competition and compensation models, not the legal status of operators per se, as often assumed in some slapdash policy prescription.

Route Allocation and Operation of Intercity Buses

In Thailand, licenses to operate intercity bus services between Bangkok and other provinces are granted solely to the state-owned operator TCL. But subcontracting is now allowed, as the underfunded state enterprise is unable to cover all areas. Private subcontractors include registered companies, limited partnerships, and individual van operators who may belong to informal operator groups called "Win" (*17*). Although Wins do not have legal status, the members may agree on and follow certain rules and codes of conduct among themselves. In recent years, the government requires that each Win set up an internal management committee in order to coordinate its operation and informally liaise with government officials in charge of van regulations. With no legally established entity, individual operators have limited resources for risk sharing, maintaining fleet quality, and acquiring vehicle insurance, let alone labor protection and social security.

As the number of informal passenger vans increased dramatically in the early 2000s, the government allowed individual van operators to become subcontractors of TCL in 2009. Such formalization drew strong criticism from bus companies, who were losing ridership to passenger vans while shouldering greater overhead costs than their informal competitors. Succumbing to pressure from the bus companies, the government allowed them to replace full-sized buses with vans for routes within 300 km at the ratio of one bus to three vans. Most private bus companies now use vans instead of full-sized buses for such short-distance routes, as the smaller vehicles are more flexible to use in response to varying demand and incur lower overhead costs. The number of intercity vans thus greatly increased, and now most routes within 300 km are operated with vans.

In terms of route allocation and operation, some routes are operated only by one operator with a legal status, which could be TCL or a private bus company. Or it could be operated only by individual informal van operators. Many other routes, however, are operated by various combinations of operator types. In the latter case, operators usually compete against one another, vying for passengers on the same route. News abounds that drivers recklessly speed and pass each other to pick up passengers before their competitors. In addition, drivers of many private bus companies and most informal passenger vans have low fixed daily or monthly salaries. Their income depends greatly on the number of trips they drive and the number of passengers on board. Such compensation levels and methods may incentivize them to drive fast and board more passengers than legally permitted, thereby increasing safety risks. By contrast, drivers of state-owned TCL buses earn fixed monthly salaries plus other benefits as state employees. Only a handful of private bus companies pay comparatively good monthly salaries and other benefits.

A bus company's de jure status does not necessarily mean the company operates as a single operational entity. Some companies or limited partnership may be registered as a single legal entity, but in practice, many of them rent out their vehicles on a daily basis to independent individual drivers, instead of operating the vehicles themselves. In other words, legal status itself may or may not translate into operational structures and procedures that lead to safety.

The vehicle types could be another source of safety risks. It is widely perceived that vans are not safe for public transport, especially when the vehicles are modified to fit in more passengers than they are designed for. As almost all bus operators who serve areas within 300 km from Bangkok now use vans, the risks could be much greater than when only full-sized buses were used. After a series of high-profile van accidents during major holidays in 2016, the government announced it would ban the use of vans and instead mandate the use of minibuses for intercity services. In addition, the government is considering banning individual operators on intercity routes and requiring that they all belong to a legal entity. If the policy is implemented, individual operators will no longer be able to acquire operating licenses from TCL.

Research Method

The key research question of this study is whether legal status affects safety risks of intercity buses. Following the government's rationale to ban individual operators on intercity bus routes, we chose the route as the unit of analysis. The main hypothesis is that the greater the number of individual operators, the greater the accident and fatality risks on a given route. Conversely, the greater the number of legally registered entities, the less the risk of accidents and fatalities on a given route.

In the regression analyses, safety risks, the dependent variable, are proxied by two measures: the number of accidents per year and the number of fatalities per year. The options for legal status, one of the key independent variables, include the state-owned TCL, private companies, limited partnerships, and individual operators without legal registration. We also control for other determinants of bus safety, including vehicle type and service standards, route characteristics (route distance and number of branches on the same route), and region of operation.

Regression Analyses

To test the hypothesis that operator's legal status affects accident and fatality risks, we use the Poisson regression model and the negative binomial regression model. We specify two groups of models with different dependent variables: one for the number of fatalities and the other for the number of accidents that occur on a given route. For the fatality risk model, let Y_i be the number of fatalities on route i, and assuming that it is distributed Poisson with the mean of μ_i. The probability mass function is given by:

$$P(Y_i = y) = \frac{\exp(-\mu_i)\mu_i^y}{y!}$$

Using the log-transformed mean, the Poisson regression model can be parameterized as follows:

$$\mu_i = \exp(X_i\beta + R_i\gamma + V_i\delta)$$

where μ_i = the expected number of fatalities (or accidents) on route i,

X_i, R_i, V_i = vectors of three groups of explanatory variables, including the number of operators in different legal status, route-specific characteristics, and the number of vehicles of different service and vehicle standards, respectively, and

β, γ, δ = vectors of coefficients for the explanatory variables.

The coefficients can be estimated straightforwardly by the maximum likelihood (ML) method. The restrictive property of the Poisson model is that the conditional mean and variance must be equal, and if this assumption does not hold, the inferences based on the standard errors of coefficients estimated by ML will be incorrect (*18*). The common violation of the Poisson assumption is the case where the variance exceeds the mean, which is known as over-dispersion. In this case, the negative binomial regression model, which allows for over-dispersion, is more appropriate (*18*), and the conditional mean and variance of Y_i can be written as:

$$\mu_i = \exp(X_i\beta + R_i\gamma + V_i\delta + \varepsilon_i)$$

and

$$Var(Y_i) = \mu_i(1 + \mu_i\alpha)$$

where $\exp(\varepsilon_i)$ = the gamma distributed error term with the mean equals 1 and the variance α.

Table 1. Route-Specific Characteristics, Operator's Legal Status, Vehicle Standards and Service Categories

Variables	Number of observations	Mean	S.D.	Min.	Max.
Number of operators with specific legal status					
Transport Company Limited	202	0.45	0.60	0	3
Private Company Limited	202	2.07	2.18	0	10
Limited Partnership	202	0.27	0.74	0	5
Independent Individual	202	16.78	30.26	0	178
Number of vehicles of different standards and service categories					
First-class air-conditioned bus (24–32 seats)	202	10.50	22.47	0	247
Second-class air-conditioned bus (40 seats)	202	2.57	6.40	0	52
Air-conditioned van (13 seats)	202	31.71	52.22	0	251
Air-conditioned double decker bus (55 seats)	202	11.73	20.01	0	110
Route-specific characteristics					
Route distance	202	402.03	284.38	39	1200
Number of branches	202	2.33	2.34	0	15

Note: S.D. = standard deviation; Min. = minimum; Max. = maximum.

When $\alpha = 0$, the negative binomial regression model reduces to the Poisson model. In this paper, we run both regressions in order to test the problem of over-dispersion in the Poisson model, and if detected, we use the robust standard errors for inferences (*19*).

We hypothesize that the occurrences of accidents and their severity as measured by the number of fatalities are different between routes of different distances due to their operational natures. As mentioned earlier, vans are allowed to operate only on the routes within 300 km. Since the amount of capital investment required for vans is substantially lower than those for full-sized buses, this may affect the organizational structures of operators, which might in turn affect safety risks. Therefore, we segment the sample and estimate separate models for short-distance (within 300 km) and long-distance (over 300 km) routes, as well as a model for all routes.

Data

The data are from two sources. First, bus route characteristics are obtained from TCL's databases, including route number, route distance, the route's geographical region, the number of branches, the number of operators of different legal status, and the number of vehicles of different service and vehicle standards. Second, bus accident characteristics are obtained from the Department of Land Transport's database, including the number of accidents, injuries, and fatalities that occurred on each route in 2015. The two data sets are merged using the route number as the key. In all, there are 202 routes, including 44, 22, 47, 33, and 56 routes between Bangkok and Central, Eastern, Northeastern, Northern, and Southern regions, respectively. Among these, 118 routes had no accident and 168 routes had no fatality in 2015.

Results

Descriptive Statistics

Route Characteristics. Table 1 provides the descriptive statistics of bus route characteristics in our data. Note that the unit of observation is the route. Each route may have more than one operator, and operators on the same route may have different legal statuses. In addition, each route may be served by more than one category of vehicles and service standards.

The most common legal status of operator is independent individual (16.78 operators per route on average) and the most common vehicle type is air-conditioned vans (31.71 vans per route on average). These two variables are highly correlated—routes with a large number of vans also have a large number of individual operators. The second most common legal status of operator is limited company, averaging just over two companies per route, while limited partnership is the least common legal status of operator, averaging only 0.27 partnerships per route.

As for vehicle standards and service categories, the double decker and first-class air-conditioned buses are second and third most common vehicle types, averaging 11.73 and 10.50 vehicles per route, respectively. The second-class 40-seat bus is the least common type, averaging only 2.57 vehicles per route. This is probably because of the competition from more efficient vans on short-distance routes and first-class buses on long-distance routes.

The average route distance is just over 402 km and the average number of branches is 2.33. Since a route may have several branch lines with different route distances, the distance used in the analysis is the average route distance of all branch lines on the same route. Note that vans are allowed to operate only on routes that are no longer than 300 km. However, some routes with an average distance of greater than 300 km may have branch lines that are shorter than 300 km, and so vans are permitted in these branch lines.

Accident Data. As shown in Figure 1*a*, the majority of routes have no accident (58.42%) and no fatality (83.17%) in 2015. One-third of all routes have one or two accidents, almost one in five routes have at least one fatality. The mean and standard deviation of the number of accidents per route are 0.64 and 1.07, respectively. The mean and standard deviation of the number of fatalities per route are 0.31 and 0.89, respectively. The distributions of route frequency by the number of accidents and fatality as well as summary statistics for routes within 300 km and routes over 300 km are provided in Figure 1, *b* and *c*, respectively.

Fatality Risk Models

As shown in Table 2, the negative binomial regression models do not fit the data very well for all specifications. The hypothesis that all of the estimated coefficients are equal to zero cannot be rejected, based on the likelihood-ratio chi-square statistics of all three models. The hypothesis of over-dispersion can be rejected for the short-distance route model, but not the long-distance route and all-route models. Therefore, for the Poisson regression model, the robust standard errors are estimated with pseudo-ML method for the latter two models and are used for inferences of their coefficients.

The Poisson models fit the data relatively better than the negative binomial models. Based on the LR chi-square statistics, the null hypothesis that all estimated coefficients are equal to zero can be rejected for all three models. The models' goodness-of-fit measures based on deviance also support the Poisson models, except for the long-distance route model, which can reject the hypothesis that the observed and the expected deviances are not different at the 0.1 significant level.

The coefficients estimate of the Poisson models show that certain variables strongly influence the number of fatalities on intercity bus routes. In the short-distance route model, two estimated coefficients are significant at the 0.01 level, that is, those for the number of individual operators and the number of vans. The coefficient for the number of individual operators is negative, implying that, all else being equal, a route with the larger number of individual operators has a lower average number of fatalities. This result is contrary to our initial hypothesis, which posits that individual operators without legal status are more prone to competition and the larger the number of this type of operators on a route should imply a higher average number of fatalities. On the other hand, the coefficient for the number of vans is positive, implying that a route with the larger number of vans has a higher average number of fatalities. This result supports our initial hypothesis that vans are less safe than buses and the larger the number of vans on a route should imply a higher average number of fatalities.

In the long-distance route models, the coefficient for the number of individual operators is positive, implying that a route with a larger number of individual operators has a higher average number of fatalities. This result is consistent with our hypothesis. The coefficient for the Central region dummy variable is negative, implying that routes in the Central region have a lower average number of fatalities, probably because there are very few long-distance routes in the Central region and these must be operated using buses only, not vans.

Accident Risk Models

Table 3 shows the regression results for the accident risk models, with the number of accidents per route as the dependent variable. Unlike the fatality risk model, the negative binomial regression models fit the data quite well for all specifications. The hypothesis that all of the estimated coefficients are equal to zero can be rejected, based on the LR chi-square statistics of all three models. The hypothesis of over-dispersion can also be rejected for all three models, allowing us to model the intercity bus accident risk and conduct statistical inference by the Poisson regression results.

Like the negative binomial models, the Poisson models of bus accident risk fit the data fairly well. The null hypothesis that all estimated coefficients are equal to zero can be rejected for all three models. Like the fatality risk model, the accident risk model's goodness-of-fit measures also support the Poisson models, except for the long-distance route model.

In the short-distance route model, the estimated coefficients for average distance and the number of vans are positive and significant at the 0.01 and 0.10 levels, respectively. These results imply that, all else being equal, the long-distance route has a higher average number of accidents, and so does the route with a larger number of vans. The types of operators have no significant impact on the average number of accidents.

In the long-distance route model, only the coefficient for the number of companies on a given route is significant at the 0.05 level. The coefficient is positive, suggesting that a route with the larger number of operators registered as company has a higher average number of accidents. The results are similar for the model for all routes, but the coefficient for the number of limited partnership is also positive and significant.

Discussion and Conclusion

Formalization of informal van services has been on the Thai government's policy agenda as part of the effort to make intercity bus services safer. The empirical results in this study categorically support the government's policy move to ban the use of vans for intercity bus services. The other key measure is to require individual operators be registered collectively as a legal entity, such as a company. The empirical results, however, do not support the government's rationale that intercity bus services are safer with more bus companies

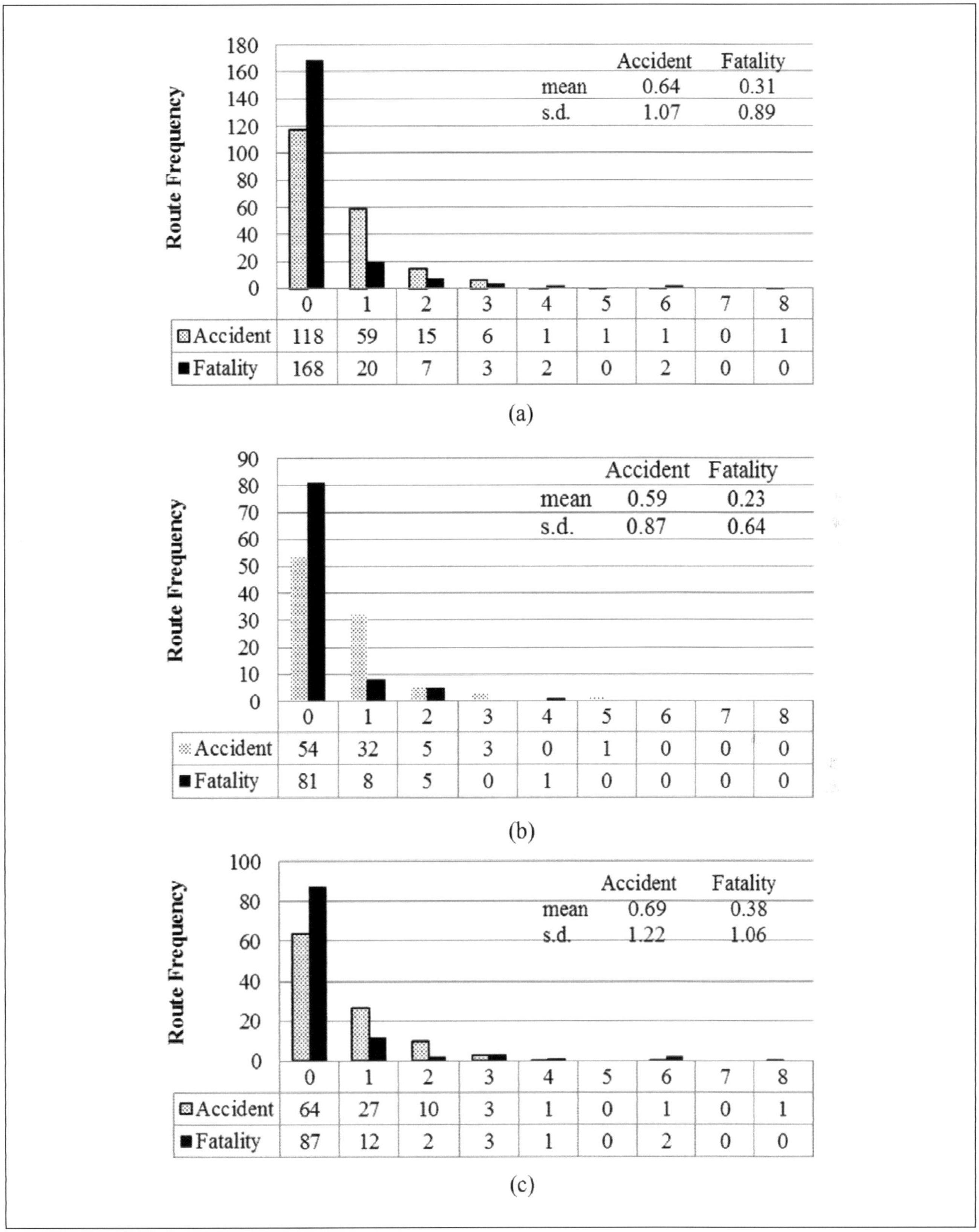

Figure 1. Frequency distribution of routes over the number of accidents and fatalities. (a) All routes; (b) Routes under 300 km; (c) Routes over 300 km.

Table 2. Regression Results: Number of Fatalities as Dependent Variable

	Poisson regression						Negative binomial regression					
	Routes under 300 km		Routes over 300 km		All routes		Routes under 300 km		Routes over 300 km		All routes	
Variables	Coefficient	s.e.	Coefficient	Robust s.e.	Coefficient	Robust s.e.	Coefficient	s.e.	Coefficient	s.e.	Coefficient	s.e.
Number of operators												
Transport Company Limited	−15.84	3678.21	0.2438	0.5130	0.4625	0.4119	−16.10	3813.35	0.3637	0.5236	0.6434	0.4156
Private Company Limited	0.1024	0.1944	0.1354	0.1233	0.1783	0.1235	0.0576	0.2167	0.1290	0.1783	0.1991*	0.1216
Partnership Limited	0.4410	0.3920	0.0362	0.3215	0.3609*	0.2206	0.5256	0.4255	−0.2041	0.8168	0.4721	0.3346
Ordinary individual	−0.0710***	0.0244	0.0832***	0.0251	−0.0181	0.0352	−0.0665**	0.0272	0.0592	0.0733	−0.0246	0.0213
Route-specific characteristics												
Average route distance (km)	−0.0015	0.0058	0.0016	0.0015	0.0004	0.0010	−0.0014	0.0062	0.0015	0.0019	0.0001	0.0011
Number of route branches	−0.0725	0.2123	−0.2456	0.2946	−0.2235	0.1687	−0.0878	0.2237	−0.0209	0.4632	−0.2369	0.1855
Central region dummy	−0.0655	0.6554	−12.0430***	1.0793	0.3682	0.7620	−0.0825	0.6818	−15.92	8148.18	0.2323	0.6941
Eastern region dummy	−0.4450	0.8341	1.6051	1.0836	0.4422	0.6470	−0.5852	0.9396	1.2630	2.1367	−0.3798	0.9361
Northern region dummy	−14.96	8828.81	0.6221	0.8076	0.2875	0.6056	−15.22	9483.47	0.5070	0.8354	0.0992	0.6663
Northeastern region dummy	73.90	3358.56	0.9000	0.7717	0.8541	0.5449	69.23	3937.98	0.8727	0.8740	0.7853	0.6668
Vehicle and service standards												
First class air-conditioned bus (24 to 32 seats)	−0.0651	0.0582	0.0125	0.0124	0.0068	0.0070	−0.0563	0.0612	0.0269	0.0277	0.0085	0.0125
Second class air-conditioned bus (40 seats)	0.0440	0.0485	0.0210	0.0503	0.0155	0.0481	0.0375	0.0553	0.0206	0.0588	0.0122	0.0376
Air-conditioned van (13 seats)	0.0401***	0.0139	−0.0125	0.0271	0.0152	0.0160	0.0385**	0.0154	−0.0126	0.0675	0.0195	0.0137
Air-conditioned double decker bus (55 seats)	−6.99	419.80	−0.0111	0.0123	−0.0135	0.0134	−6.65	492.23	−0.0152	0.0203	−0.0179	0.0165
Constant	−1.9153	0.9175	−3.2924***	1.4756	−2.2993*	0.7846	−1.8571*	0.9695	−3.6080**	1.7713	−2.1688***	0.7573
alpha	–	–	–	–	–	–	0.5086	0.9026	2.6868	1.2964	3.0152	1.1012
Number of observations	95		107		202		95		107		202	
Log likelihood	−45.5778		−81.1131		(pseudo) −143.3737		(pseudo) −45.3669		−72.3913		−125.4860	
Pseudo R-squared	0.2507		0.2158		0.1371		0.1660		0.0898		0.0687	
LR chi-squared	30.5		704		69.52		18.06		14.28		18.5	
Prob > chi-squared	0.0065		0		0		0.2041		0.4292		0.1849	
Goodness-of-fit chi-squared	58.8215		113.4428		205.6296							
Prob > chi-squared	0.9638		0.0642		0.1667							

*** 0.01 sig. level, ** 0.05 sig. level, * 0.10 sig. level

Table 3. Regression Results: Number of Accident as Dependent Variable

	Poisson regression						Negative binomial regression					
	Routes under 300 km		Routes over 300 km		All routes		Routes under 300 km		Routes over 300 km		All routes	
Variables	Coefficient	s.e.	Coefficient	s.e.	Coefficient	s.e.	Coefficient	s.e.	Coefficient	s.e.	Coefficient	s.e.
Number of operators												
Transport Company Limited	−0.4264	0.5675	−0.3147	0.2493	−0.2055	0.1951	−0.4264	0.5675	−0.2674	0.2723	−0.1847	0.2049
Private Company Limited	−0.0101	0.1062	0.1378**	0.0668	0.1076**	0.0537	−0.0101	0.1062	0.1393*	0.0808	0.1065*	0.0582
Partnership Limited	0.0228	0.1593	0.2650	0.2291	0.2580**	0.1123	0.0228	0.1593	0.2477	0.3013	0.2316*	0.1262
Ordinary individual	−0.0166	0.0116	0.0316	0.0317	−0.0080	0.0101	−0.0166	0.0116	0.0303	0.0370	−0.0085	0.0105
Route-specific characteristics												
Average route distance (km)	0.0079***	0.0026	0.0002	0.0009	0.0006	0.0005	0.0079***	0.0026	0.0002	0.0010	0.0006	0.0005
Number of route branches	0.0485	0.0840	0.0315	0.1644	−0.0245	0.0701	0.0485	0.0840	0.0833	0.2003	−0.0093	0.0737
Central region dummy	0.2927	0.3974	−13.76	959.49	0.1493	0.3345	0.2927	0.3974	−18.60	11131.37	0.1804	0.3503
Eastern region dummy	−0.1548	0.4777	0.1582	1.1696	−0.3724	0.4251	−0.1548	0.4777	0.1868	1.2708	−0.3785	0.4479
Northern region dummy	−0.5823	0.8802	−0.0904	0.3975	0.0167	0.3186	−0.5823	0.8802	−0.0673	0.4367	0.0321	0.3334
Northeastern region dummy	−0.9264	1.5408	−0.1678	0.3983	0.0223	0.2862	−0.9264	1.5408	−0.1696	0.4294	0.0068	0.3077
Vehicle and service standards												
First class air-conditioned bus (24 to 32 seats)	0.0116	0.0085	0.0097	0.0083	0.0052	0.0033	0.0116	0.0085	0.0122	0.0103	0.0062	0.0039
Second class air-conditioned bus (40 seats)	0.0011	0.0317	0.0066	0.0211	0.0067	0.0139	0.0011	0.0317	−0.0040	0.0273	0.0039	0.0148
Air-conditioned van (13 seats)	0.0124*	0.0067	−0.0125	0.0307	0.0080	0.0056	0.0124**	0.0067	−0.0163	0.0346	0.0084	0.0060
Air-conditioned double decker bus (55 seats)	0.0032	0.0319	−0.0017	0.0072	0.0056	0.0051	0.0032	0.0319	−0.0020	0.0084	0.0057	0.0059
Constant	−2.4767***	0.5851	−0.9895	0.7775	−1.2796***	0.3573	−2.4767***	0.5851	−1.0855	0.8525	−1.3362***	0.3797
alpha	–	–	–	–	–	–	0.0000	0.0001	0.2740	0.2674	0.1571	0.1629
Number of observations	95		107		202		95		107		202	
Log likelihood	−80.4930		−108.5260		−199.1030		−80.4937		−107.7300		−198.4620	
Pseudo R-squared	0.1888		0.1898		0.1476		0.1805		0.1187		0.1065	
LR chi-squared	37.46		50.86		68.98		35.46		29.03		47.32	
Prob > chi-squared	0.0006		0		0		0.0013		0.0103		0	
Goodness-of-fit chi-squared	71.4629		117.0782		208.7084							
Prob > chi-squared	0.7413		0.0399		0.1322							

*** 0.01 sig. level, ** 0.05 sig. level, * 0.10 sig. level

instead of individual operators. Quite contrary, on long-distance routes, more bus companies mean greater accident risks. One plausible explanation is that even though many operators are registered as companies or limited partnerships, they do not implement safety measures and procedures that would guarantee operational safety. Simply put, being registered as a bus company does not mean the operator has to follow more strict safety rules and operational procedures. This reflects the current regulatory loophole that business registration as company or partnership with the Ministry of Commerce does not necessarily translate into more strict regulations and oversight by the Ministry of Transport.

According to the empirical results, the only scenario where more individual operators imply higher safety risks is with fatality rates on long-distance routes. But the fatality rates are actually lower with more individual operators on short-distance routes. We conjecture that these mixed results can be attributed to the fact that most short-distance operators belong to informal operator groups, which usually have codes of conduct and internal management procedures. Such informal governance may reduce safety risks somewhat. By contrast, long-distance individual operators tend to be completely independent, so there are no internal institutions that govern their operational practice. The higher accident risks on long-distance routes also reflect the reality that the government currently does not diligently monitor and strictly enforce safety standards and procedures in intercity bus services. In conclusion, managing and monitoring service quality and safety parameters of legally registered organizations may seem easier and more effective than those of individuals. But in reality, enforcing compliance even among bus companies requires substantial institutional capacity and resources, which are currently lacking among transport agencies in Thailand.

Policy Implications: The Safety Devil is in the Detail

One could argue that legal status serves as the basis for improving driver/operator behaviors that lead to safer operation, especially when vehicle and health insurances, as well as pension and other employee benefits, are contingent on being part of a collective legal entity. But our findings suggest that legal status alone does not guarantee better safety records. That is, legalization of informal operators is not a policy panacea that would guarantee bus safety. As identified in the Haddon Matrix and numerous publications, many other things need to be implemented systematically to improve road and bus safety.

The evidence also suggests that, from the safety point of view, regulations on vehicle types could be unbundled from those on the operator's legal status and operational practices. Given that the types of van currently used in Thailand are clearly not suitable for public transport, if the government wants to allow operators to use vans, the vehicle standards have to be raised significantly. In addition, traffic monitoring and legal enforcement have to be much more strict and effective.

Future Research

This paper has several limitations. The data were at the route level, so operational characteristics that could influence drivers/operators' decisions were not analyzed. Particularly, the paper did not examine drivers' compensation levels and methods that could significantly affect incentives and behaviors of operators and drivers. A number of studies now show that shifting drivers' incentives toward safer behaviors is key to improving bus safety. Particularly important is to tie drivers' compensation with travel distance, service quality, and safety, not with the number of passengers. Therefore, future research should include a more detailed survey to acquire operator-level data that could shed light on operational and economic characteristics that may affect operators' and drivers' incentives and behaviors.

Acknowledgments

This work is the continuation of Chulalongkorn University Transportation Institute's research project, entitled "The Study of Passenger Van Safety in Public Transportation." The authors thank Chulalongkorn University's Faculty of Architecture for funding and Paniti Samnakorn for research assistance. We thank the Department of Land Transport and the Transport Company Limited for providing data for bus routes and accidents. We also thank Dr. Kasem Choocharukul for his insightful comments.

References

1. World Health Organization. *Global Status Report on Road Safety 2015*. WHO, Geneva, 2015.
2. Department of Land Transport. *Public Transport Accident Annual Report 2015*. Department of Land Transport, Bangkok, 2015.
3. Haddon, W., Jr., Advances in the Epidemiology of Injuries as a Basis for Public Policy. *Public Health Report*, Vol. 95, 1980, pp. 411–421.
4. Belzer, M. H. *The Economics of Safety: How Compensation Affects Commercial Motor Vehicle Driver Safety, October 2011*. Presented to the United States House of Representatives Committee on Small Business, July 2012.
5. Feng, S., Z. Li, Y. Ci, and G. Zhang. Risk Factors Affecting Fatal Bus Accident Severity: Their Impact on Different Types of Bus Drivers. *Accident Analysis and Prevention*, Vol. 86, 2016, pp. 29–39.
6. Wang, Y., L. Li, L. Feng, and H. Peng. Professional Drivers' Views on Risky Driving Behaviors and Accident Liability: A Questionnaire Survey in Xining, China. *Transportation Letters*, Vol. 6, No. 3, 2014, pp. 126–135.
7. Ma, M., X. Yan, H. Huang, and M. Abdel-Aty. Occupational Driver Safety of Public Transportation: Risk Perception, Attitudes, and Driving Behavior. *Transportation Research Record: Journal of the Transportation Research Board*, 2010. 2145: 72–79.
8. Cantor, D. E., H. P. Celebi, T. Corsi, and C. M. Grimm. Do Owner-Operators Pose a Safety Risk on the Nation's

Highways? *Transportation Research Part E Logistics and Transportation Review*, Vol. 59, 2013, pp. 34–47.
9. Corsi, T., C. M. Grimm, D. Cantor, and D. Sienicki. Safety Performance Differences Between Unionized and Non-Union Motor Carriers. *Transportation Research Part E Logistics and Transportation Review*, Vol. 48, No. 4, 2012, pp. 807–816.
10.Law, T. H., M. S. Daud, H. Hamid, and N. A. Haron. Development of Safety Performance Index for Intercity Buses: An Exploratory Factor Analysis. *Transport Policy*, Vol. 58, 2017, pp. 46–52.
11. Massie, D. L., and K. L. Campbell. *Analysis of Accident Rates by Age, Gender, and Time of Day Based on the 1990 Nationwide Personal Transportation Survey*. The University of Michigan, Transportation Research Institute, Ann Arbor, MI, 1993.
12. Cervero, R. *Informal Transport in the Developing World*, UN Habitat, Nairobi, 2000.
13. Govender, R., and D. Allopi. Towards a Safer Minibus Taxi Industry in South Africa. *Proc., 25th Southern African Transport Conference*, Pretoria, South Africa, 10–13 July 2006.
14. Schalekamp, H., and N. McLachlan. Minibus-Taxi Operator Reforms, Engagement and Attitudes in Cape Town. In *Paratransit in African cities: Operations, regulation and reform, Earthscan Routledge* (R. Behrens, D. McCormick, and D. Mfinanga, eds.), Oxford, 2016.
15. Estache, A., and A. Gómez-Lobo. *The Limits to Competition in Urban Bus Services in Developing Countries*. World Bank Policy Research Working Paper 3207. World Bank, Washington, D.C., 2004.
16. Dhingra, K., and S. Savant. Implications of Liberalisation of Bus Services: Case Study of Delhi. *Proc., International Conference CODATU VIII*, Cape Town, South Africa, 1998.
17. Chalermpong, S., A. Ratanawaraha, and S. Sucharitkul. Market and Institutional Characteristics of Passenger Van Services in Bangkok, Thailand. *Transportation Research Record: Journal of the Transportation Research Board*, 2016: 2581: 88–94.
18Abdulhafedh, A. Crash Frequency Analysis. *Journal of Transportation Technologies*, Vol. 6, 2016, pp. 169–180.
19. Cameron, A. C., and P. K. Trivedi. *Microeconometrics Using Stata*. Stata Press, College Station, TX, 2009.

The Standing Committee on Transportation in the Developing Countries (ABE90) peer-reviewed this paper (18-01733).

Article

TRR
JOURNAL OF THE TRANSPORTATION RESEARCH BOARD

Transportation Research Record
2018, Vol. 2672(31) 110–119

DOI: 10.1177/0361198118758638
journals.sagepub.com/home/trr
SAGE

Measuring the Passenger Car Equivalent of Small Cars and SUVs on Rainy and Sunny Days

Majid Zahiri[1] **and Xiqun (Michael) Chen**[1]

Abstract

Traffic volume is a fundamental measurement in traffic analyses. In mixed traffic, vehicles vary in size, length, headway, spacing, and acceleration/deceleration. Therefore, if we can categorize the vehicles in mixed traffic in greater detail, the estimated passenger car equivalent (PCE) number will be more accurate. Practical and appropriate methods that convert different vehicles into the equivalent number of passenger cars need to be employed to determine PCE factors for heterogeneous traffic. Following economic growth and increased use of motor vehicles in developing countries, the purchase of sports utility vehicles (SUVs) continues to grow, though the government encourages people to buy small cars because of the limited road capacity, as well as air pollution problems. In this research, we categorize passenger cars into three subsets: small cars (hatchback cars without a trunk), SUVs, and standard cars (taxis and typical family cars). A field investigation shows that the penetration rates of these passenger cars are 12%, 23%, and 55%, respectively (10% are other vehicles) in Hangzhou, China. We also measure the PCE value for small cars and SUVs using the mean time headway method. Because different countries have different weather conditions, we continue to measure the PCE values for sunny days and moderate rainy days. The results show that PCE values for small cars and SUVs are 0.87 and 1.26 on sunny days, and 0.87 and 1.31 on rainy days, respectively. By using the PCE with high precision, urban managers can accomplish the analysis of urban traffic with greater accuracy.

Traffic volume is a fundamental measurement in traffic analyses. In mixed traffic, different vehicles vary in size, length, headway, spacing, acceleration/deceleration, and so forth. A simple approach to analyzing mixed traffic is to convert various vehicles into the equivalent number of passenger cars, which is called the passenger car equivalent (PCE). There are different definitions for vehicle categories and weight classes (*1*). If it is possible to categorize vehicles in greater detail, traffic engineers will be able to analyze urban traffic with improved accuracy. Figure 1 demonstrates the schemes of an SUV, small car, and standard car. In this study, all passenger cars are classified as: (a) standard car – a regular car such as a taxi; (b) small car – a vehicle that is smaller than the standard car, such as a hatchback car without a trunk; or (c) SUV – larger and heavier than a standard car, with a higher ground clearance.

Usually, SUVs are bigger than standard passenger cars. During the 1990s, the SUV market was the fastest growing segment in the USA vehicle market, and SUV sales were about 19% of the whole light vehicle market in 1999 (*2*). In recent years, sales of SUVs in Europe have also increased. For instance, in the Netherlands, SUV sales grew significantly following the pursuit of a comfortable ride, crashworthiness, and sense of luxury (*3*). In Europe, small cars were a huge share of the market in comparison with other passenger cars; however, the trend in demand for small cars has fallen since 2009 (*4*). Chinese Automotive Fuel Economy Policy (*5*) reported that SUVs and minivans had the highest sales in the vehicle market, and multipurpose vehicle (MPVs) and SUV sales increased by 40% in 2015. In 2009, India surpassed Japan and became the larger market for small car sales at around 900,000 small cars sold compared to 700,000 sold in Japan (*6*). Nevertheless, the number of small cars in China is greater than in other countries (*5*).

Anwaar et al. (*7*) and Roess et al. (*8*) found that large vehicles influenced traffic and driver behavior in relation to physical and operational characteristics such as engine power, vehicle size, and maneuvering. A passenger car driver could not easily see the queue of vehicles ahead because of the larger size of an SUV ahead – a problem that drivers had on roads used by light-duty trucks, too (*9*). In fact, one of the effects of large vehicles on traffic is reduced sight distance. It is also clear that a longer car requires more time to pass an

[1]College of Civil Engineering and Architecture, Zhejiang University, Hangzhou, China

Corresponding Author:
Address correspondence to Xiqun (Michael) Chen: chenxiqun@zju.edu.cn

Figure 1. Schematic of an SUV (left), small car (middle), and standard car (right).

intersection and has a negative impact on the flow of traffic (*10*), meaning SUVs and small cars have different effects. By increasing or decreasing the size and mass of the vehicles, it is obvious that the space they occupy becomes greater or smaller, respectively. The time headway method used the relative amount of space occupied by a vehicle as the basis for calculating PCE values (*11*). Mardani et al. (*12*) presented that the traffic volume and composition on the road and the carriageway width of the road influenced the PCU value.

In the 1950s, Tanner (*13*) studied the influence of weather conditions on traffic flow and driver behavior. There have been many studies in this field, but to the knowledge of the authors, little attention has been paid to the effect of weather on PCE. The objective of this paper is to measure the PCE of different passenger cars in two different weather conditions (sunny days and rainy days) so that traffic engineers can analyze the effect of passenger cars on traffic parameters more precisely.

To estimate the vehicle composition of small cars and SUVs in Hangzhou, China, we collected data using cameras at nine locations. In addition, we selected three intersections in Hangzhou for data collection through video recording, and use the mean time headway method to measure the PCE of small cars and SUVs on rainy days and sunny days, respectively. We also test the randomness of the collected data using SPSS and Minitab and find the best-fitted probability distributions for different car types and study sites.

The rest of the paper is organized as follows. The next section presents a brief literature review on PCE. The third section describes the methodology to estimate PCE for small cars based on the mean time headway method, as well as the study site and data collected by cameras. The fourth section presents the time headway distributions and PCE values of small cars and SUVs, respectively. The final section concludes the paper and presents options for future research.

Literature Review

In this section, the literature on measuring PCE and the effect of weather on traffic is addressed.

Passenger Car Equivalent

Studies of PCE are important because it may affect the quality and quantity of traffic flow on roadways. The PCE concept was introduced for measuring the impact of trucks and buses on traffic problems in the *Highway Capacity Manual* (*HCM*) 1965 (*14*). Many research efforts were made to understand the impact of heavy vehicles on traffic flow (*15, 16*). Different methodological approaches have been used to find the PCE value of vehicles in previous PCE-related studies – for example, headway (*17*), density (*16*), delay (*18*), speed (*19*), travel time (*20*), volume/capacity ratio (*21*), and queue discharge (*22*). These studies mainly focused on the PCE estimation of heavy vehicles such as trucks and buses. PCE values are related to the vehicle engine power, maneuverability, vehicle weight and size, roadway construction, weather, and driver behavior; all of these factors vary from one country to another. Since small cars and SUVs are different in size, weight, engine power, maneuverability, and driver behavior, it is necessary to understand the effect of small cars, SUVs, and standard cars on traffic parameters. The knowledge of their influences on traffic is a research gap that we aim to bridge in this study.

Weather Impact

Adverse weather conditions influence driving behavior and traffic characteristics (*23*). It is clear that weather events such as rain reduce visibility and asphalt friction of the road, which have negative effects on driver behavior. There are many studies on the impact of weather conditions, but little attention is paid to the effect on PCE. According to the *HCM* 2000 (*24*), inclement weather could reduce road capacity by 15% in rainy and fog conditions. Smith et al. (*25*) classified rainy weather conditions into two groups (light rain with intensity of 0.01–0.25 inches/h and heavy rain with intensity of more than 0.25 inches/h) and found that on light rainy days and heavy rainy days, freeway capacity was reduced by 4–10 and 25–30%, respectively. Prevedouros and Chang (*26*) showed that under light rain and wet conditions, on average, there was an 8.3% reduction in freeway capacity. In the literature, to the best knowledge of the authors, there are few focused studies on the impact of weather conditions on small cars and SUVs. Thus, one of the research objectives of this study is the PCE measurement of small cars and SUVs in different weather conditions.

Figure 2. Study sites and video snapshots. (a) Site A; (b) site B; (c) site C.

PCE Estimation and Data Collection

PCE Estimation

In this study, the mean time headway method is employed for the calculation of PCE. Time headway is the time difference between two successive vehicles that pass a particular location. Greenshields et al. (*27*) estimated the PCE value using Equation 1:

$$\text{PCE} = \frac{h_{\text{small}}}{h_{\text{standard}}} \text{ or } \frac{h_{\text{SUV}}}{h_{\text{standard}}} \quad (1)$$

where h_{small} is the average headway of the small car type (it can also be replaced with other types of cars such as SUVs, i.e., h_{SUV}), and h_{standard} is the average headway of standard cars. For calculating the PCE of small cars and SUV cars, we collect the time headway of leading–following cars – that is, "small car–small car," "standard car–standard car," and "SUV–SUV" on sunny and rainy days, respectively. Usually, for measuring the PCE values of different types of vehicles, the standard passenger car is used as the basis.

Study Site Description

We select three study sites for collecting data using video cameras: a T-intersection (Yuhangtang Rd. and Zijingang Rd.); and two intersections (Wenhuai Rd. and Hedong Rd.; and Qingchun Rd. and Jianguo Middle Rd.). They are noted as sites A, B, and C, respectively. The sites are selected by considering several factors, such as saturation queues during peak hours, uninterrupted flow of traffic during the green interval (main traffic approaches), and sufficient spacing between the intersections so that the offset measurement of the same group of vehicles is avoided. In addition, the roads have no entry for trucks, no bus stops nearby, and there is no parking zone.

As shown in Figure 2, at site A we record data on two opposite approaches to the signalized T-intersection. At sites B and C, there are three and two through-lanes, respectively, in each

Table 1. The Composition of Small Cars, SUVs, and Standard Passenger Cars

Intersection #	Small cars (%)	SUVs (%)	Standard cars (%)	Others (%)
1	12	26	51	11
2	18	29	44	9
3	11	19	64	6
4	14	20	56	10
5	10	20	62	8
6	8	23	56	13
7	12	15	60	13
8	8	27	49	16
9	14	25	51	10
Mean	12	23	55	10

approach. Vehicles on the turning lanes are eliminated from this study because the driving behavior and car acceleration are different in left or right turns from the through-lanes.

Data Collection

To discover the composition of different types of cars in Hangzhou, China, we recorded videos and counted vehicles from nine typical intersections. As shown in Table 1, the average percentages of small cars, SUVs (only large and medium SUVs are included), and standard cars are 12, 23, and 55% (approximately 10% are taxis), respectively. There are 10% other vehicles such as vans. The percentages of small cars and SUVs show the importance of understanding the PCE of these vehicles and the categorization of passenger cars into subsets.

The time headway was measured between the front bumpers of two successive vehicles passing the stop line as the reference line. Weekday traffic videos were recorded for 10 h at site A and 6 h at sites B and C, respectively, during morning or evening peak hours (i.e., 8:30–10:30 a.m. or 4:30–6:30 p.m.) in sunny weather conditions on May 17, 19, and 25, 2016, and May 12 and 16, 2017, respectively. The videos were also collected in medium rainy weather conditions on December 13, 20, and 21, 2016 and May 8 and 23, 2017, respectively. The video recording was then analyzed. Time headways of small cars, SUVs, and standard cars were manually measured on a computer with a 0.01-s resolution.

Results and Discussion

In this section, we obtain the results under different weather conditions of sunny days and rainy days, which are compared to the *HCM* recommendation of PCE values.

Time Headway and PCE on Sunny Days

The data are extracted from the camera recordings of the three sites presented in above. We analyze the video and record the time headways of each pair of small cars, standard cars, and SUVs at the green light onset. To avoid the impact of fluctuations in the drivers' reaction times, we consider the time headways from the fifth vehicle in a discharging vehicular platoon, which is close to the saturate traffic flow.

Data randomness is checked with the run test of randomness by SPSS. Using this test, it is determined how far a sequence of numbers have been randomly collected at the 5% significance level. If the test value is larger than the significance level, then the random data can be verified. Table 2 shows the result of the run test of randomness for different vehicle pairs at the study sites on sunny days. All of the data significance levels (denoted by asymptotic significance) are >5%. Therefore, the observed time headways for different cars at the study sites are random.

Table 3 shows the statistics based on the probability distributions of time headway on sunny days for small cars, standard cars, and SUVs at the three study sites. We evaluate the time headway data with Minitab to determine the type of distribution. By comparing the results of different probability distribution functions, the goodness-of-fit test for each car pair is illustrated in the last two columns of Table 3. The results obtained from the probability distribution at the study sites show that the standard cars are fitted well to the distribution of three-parameter Weibull and log-logistic distributions. Weibull and log-normal distributions are the most appropriate for small cars, while log-logistic and log-normal distributions provide the best fit for SUVs. The best-fitted distributions are presented in Figure 3.

To determine the significance level of the mean time headway at different sites, we use the one-way analysis of variance (ANOVA) test by SPSS. The results are presented in Table 4. The ANOVA shows that there is a significant difference at the level of 0.99 between the different sites and the data of all sites. On the other hand, the comparison of Duncan's test for the standard car, small car, and SUV illustrates that the mean time headway of SUVs is the highest and has a significant difference to the other two vehicle types. At sites A and B, standard cars and small cars have a significant difference, but there is no significance at site C. We check all data for the three sites and the results show that there is a significant difference among these three vehicle types.

According to Equation (1), the PCE values can be determined by the mean time headway of small hatchback cars or SUVs, and standard cars. The estimated PCE values for different passenger cars are presented in Table 5. On one hand, since the small cars are smaller, and have smaller engines and less power than the passenger cars, PCE for small cars is lower than for passenger cars. The difference in the PCE value helps better characterize the mixed traffic flow. On the other hand, the PCE for SUVs is larger than for standard cars, with a mean of 1.26. An SUV may reduce the sight region for the following car driver so that the following car pays more attention to avoiding a collision. The larger size

Table 2. Run Test of Randomness for Time Headway in Sunny Days with SPSS

	Standard car			Small car			SUV		
	A	B	C	A	B	C	A	B	C
Test value[a]	2.24	2.30	2.36	1.89	2.16	2.06	3.07	2.76	2.76
Cases < test value	109	48	58	50	16	15	68	28	23
Cases ≥ test value	112	55	59	50	16	15	71	29	23
Total cases	221	103	117	100	32	30	139	57	46
Number of runs	78	41	41	47	15	15	61	32	21
Z	−4.515	−2.241	−3.435	−0.804	−0.539	−0.186	−1.613	0.671	−0.746
Asymp. sig. (2-tailed)	0.325	0.425	0.489	0.421	0.590	0.536	0.107	0.502	0.456

[a]Media.

Table 3. Time Headway Statistical Characteristics on Sunny Days

Car type and site	*n*	Mean (s)	SD (s)	Median (s)	Min (s)	Max (s)	Skewness	Kurtosis	Goodness-of-fit	*p*-value
Standard car, site A	221	2.32	0.70	2.20	1.23	4.23	0.61	−0.15	Three-parameter Weibull	0.23
Standard car, site B	103	2.37	0.74	2.30	1.10	4.40	0.46	−0.21	Three-parameter Weibull	>0.50
Standard car, site C	117	2.41	0.70	2.35	1.30	5.01	1.06	1.88	Log-logistic	>0.25
Small car, site A	100	1.90	0.40	1.88	1.24	3.25	0.74	0.92	Log-normal	0.84
Small car, site B	32	2.19	0.57	2.15	1.24	3.80	0.65	0.65	Log-normal	>0.98
Small car, site C	30	2.13	0.72	2.02	1.26	3.70	0.67	−0.39	Three-parameter Weibull	0.32
SUV, site A	139	3.07	0.91	3.09	1.21	6.26	0.67	1.65	Log-normal	<0.005
SUV, site B	57	2.89	0.81	2.77	1.31	6.10	1.33	3.91	Log-logistic	>0.25
SUV, site C	46	2.99	0.95	2.78	1.43	6.54	1.18	2.86	Log-normal	.66

Note: SD = standard deviation; SUV = sports utility vehicle.

may also result in more space between the vehicle and the following vehicle. Traffic engineers and transportation planners may benefit enormously from distinguishing PCE values for the mixed composition of vehicles in optimizing signal timing or improving road capacity, and can predict traffic characteristics with greater authenticity.

Time Headway and PCE on Rainy Days

Table 6 illustrates the results of the run test of randomness on rainy days at the same study sites as on sunny days. The asymptotic significance is also >5% for the three car types. Therefore, the time headway of the three study sites is random, too.

Table 7 illustrates the statistics based on the probability distributions of time headway on rainy days for small cars, standard cars, and SUVs at the three study sites. The best-fitted distributions are shown in Figure 4.

The results of the one-way ANOVA test for rainy days are demonstrated in Table 8. Similar to the sunny days, the level of significant difference is 0.99 and Duncan's test presents the mean time headway of SUVs as the highest and with a significant difference from the other two vehicle types. The last row of Table 8 considers all data for the three sites and shows that there is a significant difference in the mean time headway among the three vehicle types.

The estimated PCE values for different passenger cars are presented in Table 9, in accordance with Equation (1),. The mean time headway for SUVs is larger than for standard cars and small cars. On rainy days, the PCE range is 0.83–0.90 for small cars and 1.27–1.35 for SUVs.

Comparison

Comparing the mean time headways in Tables 5 and 9, it is clear that almost all cars generate greater time headway on rainy days than sunny days. For standard cars at sites A and C, there are 2.5 and 4.7% differences in the mean time headway; for small cars at the same locations these differences are 4 and 7.8% in various weather conditions. For the SUVs, the differences in the mean time headway are 1.6, 10, and 8%, respectively, at the three sites.

We find that the PCE of the small cars is less than that of the standard cars, while the PCE of the SUVs is larger than

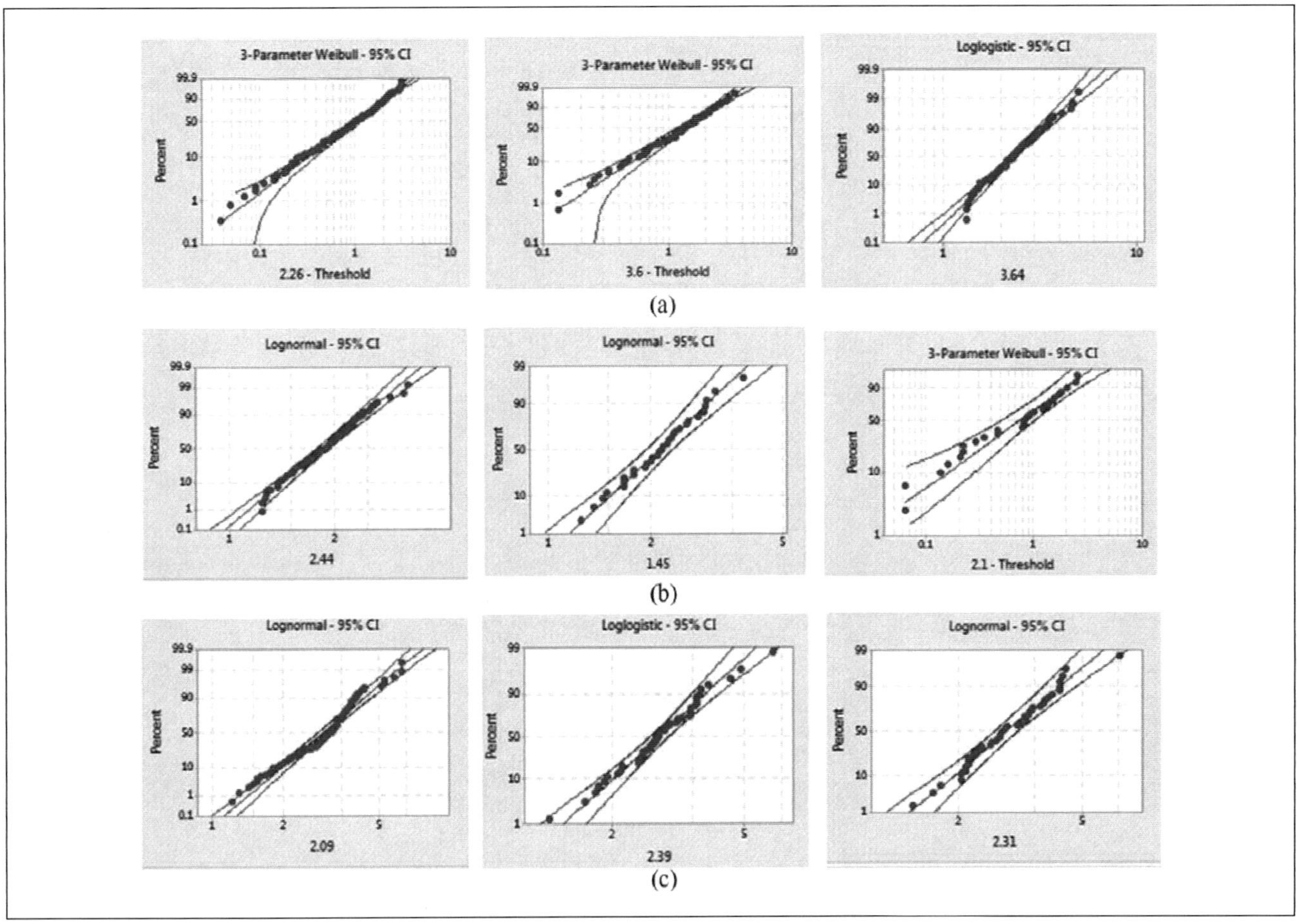

Figure 3. The best-fitted distributions of time headways on sunny days. (a) Standard car fitted distributions (left: site A, middle: site B, right: site C). (b) Small car fitted distributions (left: site A, middle: site B, right: site C). (c) SUV fitted distributions (left: site A, middle: site B, right: site C).
CI, confidence interval.

Table 4. The Analysis of Variance for Different Car Types and Comparison of Mean Time Headway by Duncan's Test on Sunny Days

Site	Standard car	Small car	SUV	SS	df	*F*-statistic	Sig.
Site A	2.32(b)	1.90(c)	3.07(a)	86.20	2	82.81	0.00***
Site B	2.37(b)	2.19(c)	2.89(a)	13.04	2	12.02	0.00***
Site C	2.41(b)	2.13(b)	2.99(a)	13.30	2	11.05	0.00***
Sites A, B, and C	2.36(b)	2.06(c)	2.98(a)	108.04	2	98.99	0.00***

Note: SS = sum of squares; df = degrees of freedom; Sig. = significance of difference between the mean time headway of different vehicles, *p-value <0.05, **p-value <0.01, ***p-value <0.001. The p-value is the probability that the test statistic can take a value greater than the value of the computed test statistic. Identical letters (a), (b), or (c) in the same row indicate that there are no significant differences between the vehicle types.

Table 5. Mean Time Headway and PCE Values on Sunny Days

Mean time headway and PCE values	Standard car	Small car	SUV
Mean time headway, site A	2.32	1.90	3.07
Mean time headway, site B	2.37	2.19	2.88
Mean time headway, site C	2.41	2.13	2.99
PCE, site A	1.00	0.82	1.32
PCE, site B	1.00	0.92	1.21
PCE, site C	1.00	0.88	1.24

Note: PCE = passenger car equivalent.

that of the standard cars. The PCE on rainy days for small cars is larger than sunny days except for site B, where the mean time headway on rainy days is larger than on sunny days. The average PCE values for the small cars across the three sites remain the same on sunny days and rainy days (0.87). The average PCE values for the SUVs on sunny days and rainy days are 1.26 and 1.31, respectively. Incorporating the difference in PCE values for weather conditions for different car sizes benefits transportation planners and traffic engineers in improving road capacity and operations. Perhaps using the weight for PCE on rainy days and sunny days

Table 6. Run Test of Randomness for Time Headway on Rainy Days with SPSS

	Standard car			Small car			SUV		
	A	B	C	A	B	C	A	B	C
Test value*	2.31	2.26	2.46	1.98	1.99	2.24	3.17	3.19	3.24
Cases < test value	102	55	53	34	15	15	57	21	21
Cases ≥ test value	103	55	53	38	15	15	58	24	21
Total cases	205	110	106	72	30	31	115	45	42
Number of runs	90	46	39	26	12	18	49	16	19
Z	−1.890	1.916	−2.928	−2.593	−1.301	0.557	−1.779	−2.090	−0.781
Asymp. Sig. (2-tailed)	0.059	0.055	0.325	0.159	0.193	0.577	0.075	0.658	0.435

*Median

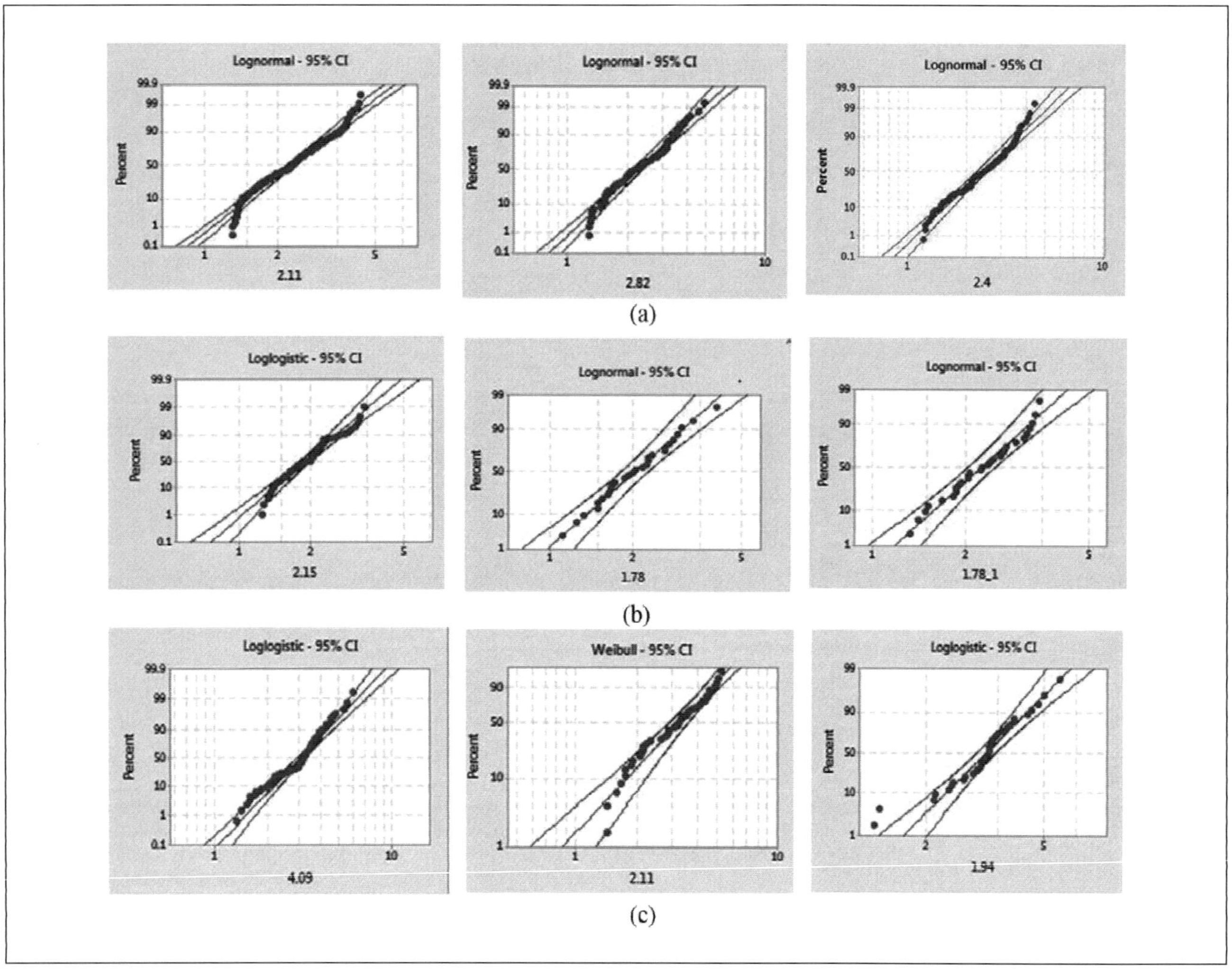

Figure 4. The best-fitted distributions of time headways on rainy days. (a) Standard car fitted distributions (left: site A, middle: site B, right: site C). (b) Small car fitted distributions (left: site A, middle: site B, right: site C). (c) SUV fitted distributions (left: site A, middle: site B, right: site C).
CI, confidence interval.

Table 7. Time Headway Statistical Characteristics on Rainy Days

Car type and site	n	Mean (s)	SD (s)	Median (s)	Min (s)	Max (s)	Skewness	Kurtosis	Goodness-of-fit	p-value
Standard car, site A	205	2.38	0.75	2.32	1.29	4.33	0.58	−0.45	Log-normal	0.007
Standard car, site B	110	2.37	0.79	2.24	1.26	4.84	0.67	−0.10	Log-normal	0.017
Standard car, site C	106	2.53	0.79	2.47	1.21	4.42	0.21	−0.83	Log-normal	0.106
Small car, site A	72	1.98	0.49	1.98	1.25	3.37	0.95	0.77	Log-logistic	0.176
Small car, site B	30	2.12	0.68	2.02	1.11	4.01	0.86	0.72	Log-normal	0.935
Small car, site C	31	2.32	0.62	2.25	1.31	3.42	0.22	−0.10	Log-normal	0.676
SUV, site A	115	3.12	0.85	3.16	1.31	5.98	0.32	0.89	Log-logistic	<0.005
SUV, site B	45	3.21	1.12	3.22	1.43	5.21	0.08	−1.18	Weibull	0.180
SUV, site C	42	3.25	0.85	3.25	1.32	5.64	0.35	1.41	Logistic	0.044

Note: SD = standard deviation; SUV = sports utility vehicle.

Table 8. The Analysis of Variance for Different Car Types and Comparison of Mean Time Headway by Duncan's Test on Rainy Days

Site	Standard car	Small car	SUV	SS	df	F-statistic	Sig.
Site A	2.38(b)	1.98(c)	3.12(a)	68.07	2	61.73	0.00***
Site B	2.37(b)	2.12(b)	3.21(a)	27.82	2	18.67	0.00***
Site C	2.53(b)	2.32(c)	3.25(a)	18.99	2	15.55	0.00***
Sites A, B, and C	2.43(b)	2.13(c)	3.19(a)	112.72	2	91.83	0.00***

Note: SS = sum of squares; df = degrees of freedom; Sig. = significance of difference between the mean time headway of different vehicles, *p-value <0.05, **p-value <0.01, ***p-value <0.001. The p-value is the probability that the test statistic can take a value greater than the value of the computed test statistic. Identical letters (a), (b), or (c) in the same row indicate that there are no significant differences between the vehicle types.

Table 9. Mean Time Headway and PCE Values on Rainy Days

Mean time headway and PCE values	Standard car	Small car	SUV
Mean time headway, site A	2.38	1.98	3.12
Mean time headway, site B	2.37	2.12	3.21
Mean time headway, site C	2.53	2.31	3.25
PCE, site A	1.00	0.83	1.31
PCE, site B	1.00	0.89	1.35
PCE, site C	1.00	0.90	1.27

Note: PCE = passenger car equivalent.

according to the number of rainy days and sunny days for the whole year is a solution for estimating long-term localized PCE. For example, there are about 60 rainy days in each year in Hangzhou, so we have estimated PCE for SUVs given by $PCE = (1.26 \times 305 + 1.31 \times 60)/365 = 1.27$. The *HCM* 2010 recommends that the PCE for passenger cars equals 1. As shown in Table 10, we divide the passenger cars into the three groups for localized PCE estimation.

Conclusions

PCE is defined as a dimensionless number based on which a certain vehicle can be converted into a standard vehicle for traffic analyses. PCE depends on the car size, engine power, road type, and driving behavior. In this study, for understanding traffic characteristics in mixed traffic with greater accuracy, we classified the passenger cars into three subset groups: small cars, SUVs, and standard cars. We measured the PCE of these three groups on rainy days and sunny days using the mean time headway method. We also investigated the percentages of these three groups by video cameras in Hangzhou; these were 12% (small cars), 23% (SUVs), and 55% (standard cars). Based on the empirical data collection, the results of this paper recognize that the PCEs of small cars and SUVs are not the same or equal to 1, as recommended in the *HCM*. The PCE of small cars is 0.87 on both sunny and rainy days, and the PCE for SUVs is 1.26 on sunny days and 1.31 on rainy days. It shows that there is not a significant difference between the small car PCE on sunny days and rainy days, but clarifies that small cars have about 0.13 smaller PCE than standard cars. SUVs have 0.26 and 0.31 greater PCE than standard cars on sunny days and rainy days, respectively.

One limitation of this research is that there are only a few small cars observed in the study sites, thus the number of small cars recorded is lower than the SUV and standard cars. The PCE measurement of small cars and SUVs on rainy and sunny days can be expanded in future research: (a) calculating the delay induced by small cars, SUVs, and passengers cars; (b) determining the PCE of small cars and SUVs on freeways under various weather conditions; and (c) understanding the impact of small cars and SUVs on road capacity.

Table 10. Comparison of the Car-Specific PCE with HCM Recommendations

PCE and traffic composition	Small cars	Standard cars: Private cars	Standard cars: Taxis	SUVs
PCE of HCM recommendation	1.00	1.00	NA	1.00
Average PCE on sunny days (this study)	0.87	1.00	NA	1.26
Average PCE on rainy days (this study)	0.87	1.00	NA	1.31
Traffic composition (this study)	13.3%	50%	11.1%	25.6%

Note: NA = not available; PCE = passenger car equivalent. According to (*28*), taxi PCE is usually >1.

Acknowledgments

This research is financially supported by Zhejiang Provincial Natural Science Foundation of China [LR17E080002], National Natural Science Foundation of China [51508505, 71771198, 51338008], and Fundamental Research Funds for the Central Universities [2017QNA4025].

References

1. An, F., R. Earley, and L. Green-Weiskel. *Global Overview on Fuel Efficiency and Motor Vehicle Emission Standards: Policy Options and Perspectives for International Cooperation.* United Nations, Department of Economic and Social Affairs, Background Paper No. 3, CSD19/2011/BP3, New York, May 2–13, 2011.
2. Davis, S. C., and L. F. Truett. *An Analysis of the Impact of Sport Utility Vehicles in the United States.* Oak Ridge National Laboratory, No. ORNL/TM-2000-147, Oak Ridge, TN, 2000.
3. Margaritis, D., B. Hoogvelt, Y. de Vries, C. Klootwijk, and H. Mooi. *An Analysis of Sport Utility Vehicles Involved in Road Accidents.* Enhanced Safety of Vehicles, Washington, D.C., 2005.
4. The International Council on Clean Transportation. *European Vehicle Market Statistics: Pocketbook.* The International Council on Clean Transportation, 2015. http://eupocketbook.theicct.org. Accessed on December 1, 2016.
5. The Chinese Automotive Fuel Economy Policy. https://www.globalfueleconomy.org/transport/gfei/autotool/case_studies/apacific/china/CHINA%20CASE%20STUDY.pdf. Accessed March 30, 2018.
6. Ghosh, A., S. Ray, and S. Dewan. Indian Passenger Vehicle Industry: Growth Momentum to Continue. In *Indian Passenger Vehicle Industry: An ICRA Perspective*, ICRA Limited, March 2011, pp. 1–15. New Delhi, India.
7. Anwaar, A., D. Van Boxel, M. Volovski, P. C. Anastasopoulos, S. Labi, and K. C. Sinha. Using Lagging Headways to Estimate Passenger Car Equivalents on Basic Freeway Sections. *Journal of Transportation of the Institute of Transportation Engineers*, Vol. 2, No. 1, 2011, pp. 5–17.
8. Roess, R. P., and E. S. Prassas. *The Highway Capacity Manual: A Conceptual and Research History.* Springer, New York, 2014.
9. Shabih, R. A. *Effect of Vehicle Type on the Capacity of Signalized Intersections: The Case of Light-Duty Trucks.* Doctoral dissertation, University of Texas, Austin, 1999.
10. Kockelman, K. M., and R. A. Shabih. Effect of Light-Duty Trucks on the Capacity of Signalized Intersections. *Journal of Transportation Engineering*, Vol. 126, No. 6, 2000, pp. 506–512.
11. Cunagin, W. D., and C. J. Messer. *Passenger Car Equivalents for Rural Highways.* Federal Highway Administration, No. FHWA-RD-82-132, 1982. Washington, DC.
12. Mardani, M. N., S. Chandra, and I. Ghosh. Passenger Car Unit of Vehicles on Undivided Intercity Roads in India. *Procedia Computer Science*, Vol. 52, 2015, pp. 926–931.
13. Tanner, J. C. Effect of Weather on Traffic Flow. *Nature*, Vol. 169, No. 4290, 1952, pp. 107–107.
14. *Special Report 87: Highway Capacity Manual*, 2nd ed. HRB, National Research Council, Washington, D.C., 1965.
15. Sumner, R., D. Hill, and S. Shapiro. Segment Passenger Car Equivalent Values for Cost Allocation on Urban Arterial Roads. *Transportation Research Part A*, Vol. 18, No. 5–6, 1984, pp. 399–406.
16. Webster, N., and L. Elefteriadou. A Simulation Study of Truck Passenger Car Equivalents (PCE) on Basic Freeway Sections. *Transportation Research Part B*, Vol. 33, No. 5, 1999, pp. 323–336.
17. Krammes, R. A., and K. W. Crowley. Passenger Car Equivalents for Trucks on Level Freeway Segments. *Transportation Research Record: Journal of the Transportation Research Board*, 1986. 1091: 10–17.
18. Craus, J., A. Polus, and I. Grinberg. A Revised Method for the Determination of Passenger Car Equivalencies. *Transportation Research Part A*, Vol. 14, No. 4, 1980, pp. 241–246.
19. Elefteriadou, L., D. Torbic, and N. Webster. Development of Passenger Car Equivalents for Freeways, Two-Lane Highways, and Arterials. *Transportation Research Record: Journal of the Transportation Research Board*, 1986. 1572: 51–58.
20. Keller, E. L., and J. G. Saklas. Passenger Car Equivalents from Network Simulation. *Journal of Transportation Engineering*, Vol. 110, No. 4, 1984, pp. 397–411.
21. Linzer, E. M., R. P. Roess, and W. R. McShane. Effect of Trucks, Buses, and Recreational Vehicles on Freeway Capacity and Service Volume. *Transportation Research Record: Journal of the Transportation Research Board*, 1979. 699: 17–26.
22. Al-Kaisy, A., Y. Jung, and H. Rakha. Developing Passenger Car Equivalency Factors for Heavy Vehicles during Congestion. *Journal of Transportation Engineering*, Vol. 131, No. 7, 2005, pp. 514–523.
23. Akin, D., V. P. Sisiopiku, and A. Skabardonis. Impacts of Weather on Traffic Flow Characteristics of Urban Freeways in Istanbul. *Procedia-Social and Behavioral Sciences*, Vol. 16, 2011, pp. 89–99.
24. *Highway Capacity Manual*. TRB, National Research Council, Washington, D.C., 2000.

25. Smith, B. L., K. G. Byrne, R. B. Copperman, S. M. Hennessy, and N. J. Goodall. An Investigation into the Impact of Rainfall on Freeway Traffic Flow. Presented at 83rd Annual Meeting of the Transportation Research Board, Washington, D.C., 2004.
26. Prevedouros, P. D., and K. Chang. Potential Effects of Wet Conditions on Signalized Intersection LOS. *Journal of Transportation Engineering*, Vol. 131, No. 12, 2005, pp. 898–903.
27. Greenshields, B. D., D. Schapiro, and E. L. Ericksen. *Traffic Performance at Urban Street Intersections*. Technical Report 1, 1946. Eno Foundation for Highway Traffic Control, Washington, DC.
28. Golias, J. C. Examining Sensitivity of Impact of Taxi Traffic on Capacity and Delays at Urban Road Sections. *Journal of Transportation Engineering*, Vol. 129, No. 3, 2003, pp. 286–291.

The Standing Committee on Transportation in the Developing Countries (ABE90) peer-reviewed this paper (18-04359).

Article

TRR
JOURNAL OF THE TRANSPORTATION RESEARCH BOARD

Transportation Research Record
2018, Vol. 2672(31) 120–129

DOI: 10.1177/0361198118758334
journals.sagepub.com/home/trr
SAGE

Impact of Curbside Parking on Bicycle Lane Capacity in Nanjing, China

Xiaofei Ye[1], Xingchen Yan[2], Jun Chen[2], Tao Wang[2], and Zhao Yang[3]

Abstract

As roadway resources are being occupied by curbside parking and because of the operational characteristics of parking maneuvers, the capacity of the adjacent travel lane can be significantly reduced. To analyze the influence of curbside parking on the capacity of the bicycle lane, a conflict technique based on additive conflict flow was applied to establish the base capacity model. The actual capacity of the bicycle lane with curb parking was then established by adjusting the base capacity to reflect the influence: lane width, the time influence of parking maneuvers, and proportion of e-bikes. Eight datasets from the exclusive bicycle lanes with different widths and parking maneuvers were collected in Nanjing, China for calibration and evaluation purposes. As a result of a higher number of parking maneuvers, the Emeiling Road was taken as the main case study. The capacity of the bicycle lane was calculated, and the effectiveness of the proposed method was validated by the speed-density-volume relationship model. The proposed model was applied to analyze the effect of different positions of parking berths on the capacity. The results indicate that, with around 65% share of e-bikes, the estimated capacity of Emeiling Road is 2622 bicycles/h, decreasing by 47.10% under the influence of curbside parking. The results also imply that the berths near the openings of the isolation belt have less influence than those in the middle position. These findings could be helpful and useful for practitioners to improve the capacity of bicycle lanes under the influence of curbside parking.

As an important part of the parking system, curbside parking refers to using the road's spatial resources (i.e., hard shoulders or wide lanes) as the parking carrier. Compared with off-street parking facilities, curbside parking is more flexible, space-saving, and convenient in contrast to off-street parking facilities. Conversely, traffic engineers advocate that curbside parking should be prohibited. They point out that curbside parking occupies the roadway resources that directly influence traffic performance on the adjacent travel lane and that the parking maneuvers (entering and exiting parking stalls) increase traffic delays and reduce the capacity of the adjacent travel lane. Additionally, curbside parking has also been found to increase traffic accidents (*1*).

The merits and drawbacks of curbside parking have made it a controversial topic. In China, it has been suggested to place curbside parking on branch roads with low volume, but more and more exclusive bicycle lanes with physical separation between motorized and non-motorized vehicles (shown in Figure 1), and a high volume of non-motorized vehicles have been permitted to use curbside parking stalls. Although exclusive bicycle facilities with curbside parking are a specific type of road design which is of a limited application in the United States and European countries, there is a wide range of applications in China. This is a controversial phenomenon, because bicycles and electric-bikes (e-bikes) are directly influenced by the parking stalls. Therefore, a rigorous influential analysis of curbside parking is necessary for its design and optimization. Additionally, there have been relatively few studies on the bicycle capacity in developed countries because of the small volume of bicycles and there are still some questions as to whether the existing methods are applicable to the high bicycle volume conditions and heterogeneous traffic comprising of e-bikes and regular bicycles in China. Therefore, the purpose of this paper is to quantify the influence of curbside parking on the capacity of the bicycle lane, and find out how the factors of curbside parking reduce the bicycle lane capacity.

In this paper, a base capacity model derived by a traffic technique is established by considering the influence of curbside parking. The actual capacity for the bicycle lane with curb parking is then established by adjusting the base

[1]Ningbo Collaborative Innovation Center for Port Trade Cooperation and Development, School of Maritime and Transportation, Ningbo University, Ningbo, China
[2]School of Transportation, Southeast University, Nanjing, China
[3]College of Civil Aviation, Nanjing University of Aeronautics and Astronautics, Nanjing, China

Corresponding Author:
Address correspondence to Xiaofei Ye: yexiaofei@nbu.edu.cn

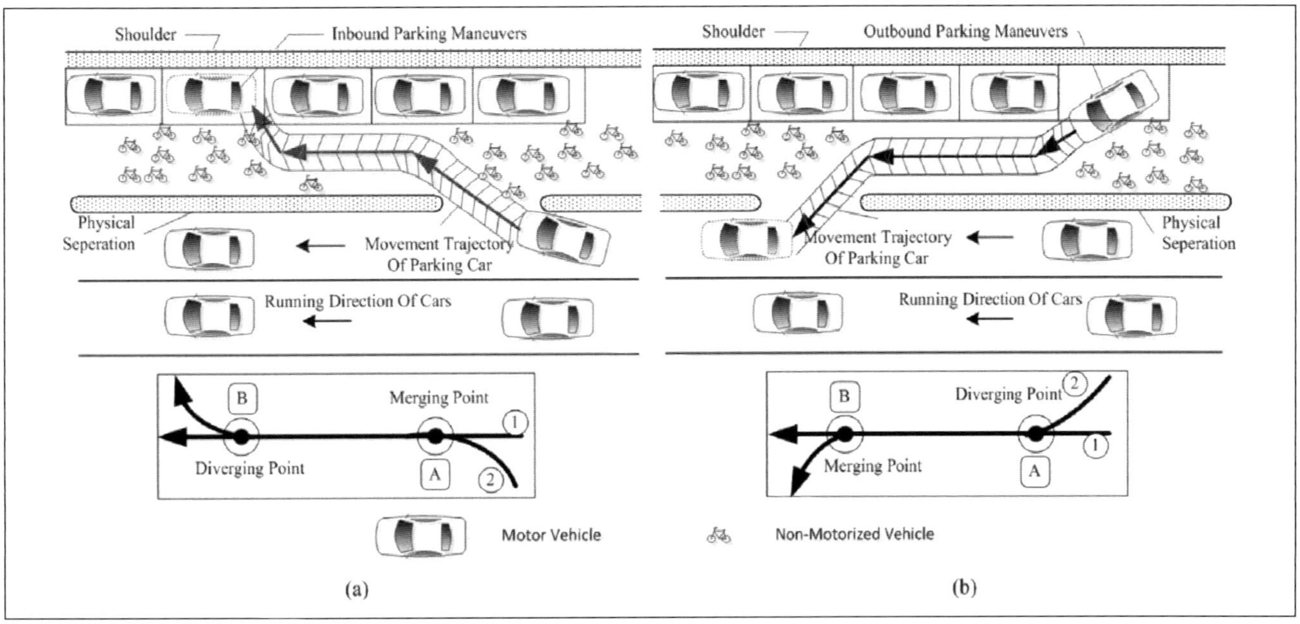

Figure 1. Conflicts between non-motorized vehicles and parking cars, (*a*) conflict procedures caused by inbound parking maneuvers, and (*b*) conflict procedures caused by outbound parking maneuvers.

capacity to reflect the influence: lane width, the time influence of parking maneuvers, and the proportion of e-bikes. The empirical data, which were obtained from volume-speed survey data on the bicycle lane with curbside parking in Nanjing, China, are prepared for model calibration. Finally, the capacity of the bicycle lane is calculated, and the proposed model is applied for discussing the effect of parking maneuvers and positions on the bicycle lane capacity.

Literature Review

Influence of Curbside Parking

Chick and the Highway Capacity Manual took the loss of travel lane width and the added friction of the number of parking maneuvers per hour as the variables to quantify the impact of curbside parking bordering the travel lane on reduced capacity (*1,2*). Valleley also considered the number of parked cars and the loss of adjacent travel lane width as influential factors and proposed a relationship model of capacity (*3*). Yousif established a delay model influenced by the number of inbound and outbound parking maneuvers considering parallel and angle parking types (4). Arnott focused on the cruising behavior of a parker for a vacant parking stall and established a parking congestion model (*5*). Marshall reassessed the different aspects of curbside parking, such as parking demand and supply, land use and the influence on travel velocity (*6*). Portilla established the travel time model by using the M/M/∞ queuing theory and considering the influence of parking maneuvers (*7*). Additionally, it has been documented in the literature that curbside parking increases traffic accidents (*8*).

As described above, the previous research in this area has focused on the influence of curbside parking on motorized vehicles. However, non-motorized vehicles are directly influenced by parking and little research has been performed that considers the effect on bicycles and e-bikes. Ye was the first to propose a proportional hazard-based duration model for quantifying the influence of curbside parking on the travel speed of non-motorized vehicles (*9*).

Bicycle Lane Capacity

The main method of capacity estimation is to calculate the maximum traffic flow in saturated conditions. Saturation flow can rarely be observed on bicycle facilities in the United States and European countries, so the capacity is always calculated by sparse data from China, other Asian countries or simulations. The Highway Capacity Manual lists ranges of reported capacities for bicycle facilities between 500 and 2,350 bicycles/h depending on the type of facility, the number of lanes, and whether it is one-way or two-way operation (*2*). Homburger recommends that bicycle facility capacity can reach approximately 2,600 bicycles/h/m (*10*). Ferrara also proposed the simulation model for bicycle lane widths of 1.2 m, 1.8 m, and 2.4 m (*11*). The study used the data gathered from the simulation model to predict the saturation flow. The results show that the saturation flow for a 1.2 m lane width reached 3,060 bicycles/h. The Swedish Capacity Manual recommends 1,500 bicycles/h as the planning capacity of a 1.2 m bicycle lane (*12*). A Canadian study showed that the capacity of a 2.5 m bicycle path is 10,000 bicycles/h (*13*). Liu used the data collected from Beijing, China in 1993, for a bicycle capacity model, and found capacities of between

1,800 and 2,100 bicycles/h per 1 m lane (*14*). Botma concluded that the bicycle capacity of a 2 m two-lane path was 6,400 bicycles/h and the capacity of a 3 m three-lane path was 9,600 bicycles/h in the Netherlands (*15*). However, another Dutch study reported bicycle capacities of 3,000 to 3,500 bicycles/h per 0.78 m of bicycle lane (*16*). Wei observed bicycle flow operation in Beijing and proposed the massive bicycle flow theory (*17*). The capacity of uninterrupted bicycle facilities with physical separation between non-motorized and motorized vehicles is 2,344 bicycles/h. This is slightly lower than the reported theoretical capacity of 2,549 bicycles/h calculated using the characteristics of individual Chinese bicyclists. The Ministry of Housing and Urban–Rural Development of China recommended that the capacity of the bicycle lane with purely regular bicycles, was 1600–1800 bicycles/h/m with the physical separation and 1400–1600 bicycles/h/m without the physical separation (*18*). Additionally, some research has considered the bicycle lane capacity for a mixture of regular bicycles and e-bikes. Jin reports the capacity of a bicycle lane, with around a 70% proportion of e-bikes is 2,348 bicycle/h/m, and with purely regular bicycles and purely e-bikes is found to be 1800 and 2727 bicycles/h/m, respectively (*19*).

In summary, the capacity for a single 1.0 m to 1.2 m bicycle lane appears to be between 1,500 and 5,000 bicycles/h and differs greatly from different studies in different countries. The majority of the observed capacities fall between 2,000 and 3,500 bicycles/h. Most of the previous research focuses on the bicycle lane capacity for uninterrupted flow under prevailing roadway, traffic, and control conditions. Little research has been conducted to study the capacity when there is a curbside parking setting on the bicycle lane. In addition, how the curbside parking affects the capacity needs to be modeled, analyzed and validated. A good understanding of the influence of curbside parking is the basis of the design and optimization of both curbside parking and bicycle facilities.

Methodology

The derivation is started by looking at the conflict between parking cars and non-motorized vehicles. When a parking car enters or exits a parking stall, the conflict occurs because of the parking maneuvers. Conflict arises from the segment of two movements that must pass through the same area. Consequently, the parking car and non-motorized vehicles involved in a conflict must pass through the area one after the other. The two movements involved in a particular conflict are defined as a "conflict group." Although parking vehicles have a higher priority than non-motorized vehicles because of the continuous occupancy of parking maneuvers on the bicycle lane.

To describe this conflictual phenomenon, a conflict technique based on the additive conflict flows (ACFs) is applied to analyze the impact of curbside parking on the bicycle capacity. The ACFs-based modeling has been applied extensively to describe the conflicts among different steams and calculate the capacity in two-way stop-controlled, all-way stop-controlled, and uncontrolled intersections (*20, 21*). As an analogy, the concept of ACFs will be developed for capacity estimation of the bicycle lane with curbside parking.

Capacity Model of Bicycle Lane Influenced by Curbside Parking

Base Capacity Derived by Conflict Technique. As shown in Figure 1, a conflict group is generated in the same conflict area when a parking car enters or exits a parking stall. Each conflict group involves three conflict points consisting of a merging point, diverging point and "fluctuant follow-up" conflict section. The conflict area can only be occupied by either a parking car or non-motorized vehicles at one time. One conflict group usually contains inbound or outbound parking vehicular movements and non-motorized vehicles.

According to ACFs-based model of an uncontrolled intersection, it is assumed that the non-motorized vehicles movement ($i = 1$) occupies the conflict area for the exact time t_{Bi}. The total time available for non-motorized vehicles and parking cars from two movements is 3,600 s/h.

If there are no parking cars entering or exiting a parking stall, the non-motorized vehicles can pass through the conflict area without the influence of parking cars. The capacity of non-motorized vehicles, C_0, is the maximum number of non-motorized vehicles that can pass through the conflict area without any influence:

$$C_0 = \frac{3600}{t_{B1}} \tag{1}$$

In which C_0 denotes the maximum capacity of non-motorized vehicles without the influence of parking cars and t_{B1} is the average time for non-motorized vehicles crossing the conflict areas and is equal to the safe headway of non-motorized flow without the interference by other movements.

If there are conflicts between the parking cars and non-motorized vehicles, they have to pass through the conflict area alternately. Two movements must have the same capacity in arrival sequence if the traffic flow Q_i exceed the capacities C_i (total overload). The capacities of all movements have the same value under oversaturated conditions:

$$C_i = \frac{3600}{\sum t_{Bi}} = \frac{3600}{t_{B1} + t_{B2}} \text{ for } Q_i > C_i \tag{2}$$

In which t_{B2} is service time for a parking car crossing the conflict areas.

However, parking cars usually occupy the conflict area for a moment to complete the parking maneuvers and do not consume the admitted capacity, and the remaining capacity can then be used by non-motorized vehicles. In this case, for

$Q_2 < C_2$, the capacity of the non-motorized vehicles can be expressed as follows:

$$C_1 = \frac{3600 - Q_2 \cdot t_{B2}}{t_{B1}} \tag{3}$$

To summarize, let P_c denotes the probability that the conflict area is occupied by parking cars and non-motorized vehicles simultaneously. The base capacity of the bicycle lane influenced by curbside parking C_p can be expressed as follows:

$$C_p = C_0 \bullet (1 - P_c) + C_1 \bullet P_c \tag{4}$$

In which P_c can be determined by the conflict probability model in the next section.

Determining the Conflict Probability between Parking Cars and Non-Motorized Vehicles. There are four major types of parking events consisting of inbound and outbound parking maneuvers, double parking and load/unload maneuvers. Different parking maneuvers occupy different conflict areas and have different influence time durations. As double parked vehicles and load/unload maneuvers rarely happened in the survey (only twice), they were removed. Let m_i, n_i denote the number of inbound and outbound parking maneuvers at the berth *i*, respectively, the total service time of parking cars in an hour can be calculated by the following equation.

$$Q_2 t_{B2} = \sum_{i=1}^{M} m_i t_{B2}^{in} + n_i t_{B2}^{out} \tag{5}$$

In which Q_2 is defined as counts of inbound and outbound parking cars at all the berths *i*, $Q_2 = \sum_{i=1}^{M} m_i + n_i$, M is the total parking berths, t_{B2}^{in} and t_{B2}^{out} denote the service time for inbound and outbound parking cars crossing the conflict areas, respectively.

Considering the independence of inbound and outbound parking events, the mean conflict probability caused by curbside parking $\bar{P}_C$ can be written as:

$$\bar{P}_C = \frac{\sum_{i=1}^{M} m_i \bullet P_{ci}^{in} + \sum_{i=1}^{M} n_i \bullet P_{ci}^{out}}{\sum_{i=1}^{M} m_i + \sum_{i=1}^{M} n_i} \tag{6}$$

In which P_{ci}^{in} and P_{ci}^{out} are conflict probabilities caused by inbound and outbound parking cars, respectively, at the berth *i*.

There are two conditions that generate conflicts between parking cars and non-motorized vehicles:

1. When the parking cars enter or exit the parking stall and occupy the conflict areas, the non-motorized vehicles arrive at the conflict area simultaneously;
2. The headways of the massive non-motorized vehicles are not distinctly greater than the critical gap t_c required by parking maneuvers.

The conflict probability that two conditions are met can be written as:

$$P_c = P(t \le \min\{T, t_c\}) = \int_0^{\min\{T, t_c\}} f(t)dt \tag{7}$$

In which P_c denotes the conflict probability between parking cars and non-motorized vehicles, *t* is the headway of the non-motorized traffic flow, *f*(*t*) denotes the probability density function of *t*, and *T* is the service time for parking car occupying the conflict area.

Determining the Adjustment Factors of Bicycle Lane Capacity

The actual capacity of the bicycle lane is defined as a result of basic or ideal capacity with various adjustment factors and corrections which reflect the effects of geometric, environmental and traffic conditions. The actual capacity can be calculated as:

$$C = C_p \times f_w \times f_T \times f_{eb} \times f_j \tag{8}$$

In which C is the actual capacity of the bicycle lane with curbside parking, C_p is the base capacity, f_w is adjustment factor for the effective width of the bicycle lane, f_T is the adjustment factor for the time influence rate of curbside parking, f_{eb} is the e-bike adjustment factor, and f_j is comprehensive correction coefficient combined intersection with pedestrians, with the suggested default value of 0.55 (*18*).

Adjustment for bicycle lane width. There are no explicit driveway divisions for the bicycle lane like in a vehicular roadway. The minimum width of a single bicycle lane can be determined by the size of different types of bicycles and e-bikes. The size datasets were collected in Nanjing, 400 bicycles and e-bikes were observed. The widths of 85% of bicycles and e-bikes are 0.60 and 0.68 m, respectively. The width for a single bicycle lane was calculated by the sum of the width of bicycles and e-bikes, and the safety lateral clearance (0.25 m) was 1.10 and 1.18 m. Compared with other research, the single bicycle lane width b_e is recommended as 1.25 m. Adjustment to reflect the effect of bicycle lane width is given as:

$$f_w = \frac{W_e + 0.5}{b_e} = \frac{W_e + 0.5}{1.25} \tag{9}$$

In which W_e denotes the effective width of the lane that reflected the reduction in the lateral clearance caused by curbside parking spaces.

Adjustment for the Time Influence Rate of Curbside Parking. The time influence rate is defined as the ratio of the total influence time T to the statistical time interval T_d (in hours). It is described as the duration of a parking maneuver. The adjustment to reflect the effect of the time influence rate is provided as:

Table 1. Characteristics of Test Sections

Test section	Effective width (m)	Parking maneuvers (events/h)	Time influence rate (%)	E-bike volume (e-bikes/h)	Bicycle volume (bicycles/h)	Average percentage of e-bikes	Average speed of nonmotorized vehicles (m/s)	Average speed of parking vehicles (m/s)
Zhongshan North Road A	3.80	63	12.41	864	537	0.64	4.73	4.11
Zhongshan North Road B	4.50	41	14.90	1529	872	0.63	5.47	4.67
Zhongshan North Road C	3.0	51	16.80	1500	800	0.65	4.25	3.63
Emeiling Road	3.50	91	21.75	1102	588	0.65	4.41	3.72
Shuiximen Road	2.20	46	10.68	1004	670	0.61	3.45	2.75
Guangzhou Road	3.40	48	13.76	906	446	0.67	4.35	3.68
Huaqiao Road	3.20	57	17.12	1243	585	0.68	4.31	3.7
Shanghai Road	2.60	21	11.45	876	661	0.57	3.86	2.95

$$f_T = 1 - \frac{\sum_{j=1}^{N}\left(t_3^j - \frac{L_{n(j)}}{V}\right)}{3600} \quad (10)$$

In which $L_{n(i)}$ denotes the distance of the follow-up conflict area between the merging point and diverging point, V is the free-flow speed of non-motorized vehicles, N is the sums of inbound and outbound parking maneuvers, and t_3^j is service time for parking car j crossing the follow-up conflict areas, which is given as:

$$t_3^j = \frac{L_{n(j)}}{V_1} \quad (11)$$

In which V_1 is the actual speed of parking vehicles.

Electric Bike Adjustment Factor. E-bikes are an important component of non-motorized vehicles. They account for about 65% of the non-motorized vehicles in the survey. In an analogy to heavy-vehicle adjustments of vehicular flows, e-bikes must be adjusted to an equivalent flow rate expressed in bicycles per hour per lane. The adjustment factor f_{eb} is determined by using equation:

$$f_{eb} = \frac{1}{1 + eb_i(n_{ec} - 1)} \quad (12)$$

In which eb_i denotes the proportion of e-bikes in the non-motorized vehicles, and n_{ec} is the bicycle equivalent units for e-bike that converts the e-bike into a standard or basic bicycle. The suggested default value is 1.23 (*22*).

Data Collection

Volume-Speed Survey Considering Curbside Parking

The field data are video recordings which were collected during peak hour (7:00.–9:00 a.m. or 4:30.–7:00 p.m.) in Nanjing, China. The following criteria were used to choose appropriate urban streets for data collection:

1. On all of the study sites selected there exists a bicycle lane with curbside parking and physical separation between motorized and non-motorized vehicles. Additionally, all bicycle lanes are usually one-way.
2. Non-motorized vehicles under isolated parking maneuver conditions are considered in the study. In some cases the non-motorized vehicles affected by the events other than parking maneuvers, double parking or load/unload condition maneuvers are excluded.

The following general procedures were implemented on the data reduction; firstly, all the bicycles and e-bikes passing through the street segment were counted. From this, the flow for a one minute interval was calculated. Secondly, the average travel time including stops and delay in each one minute time interval was calculated, the duration of parking maneuvers was recorded and then the travel speed of parking cars and non-motorized vehicles was calculated using the length of the street segment. Meanwhile, parking maneuvers were summed up under each category; the inbound and outbound maneuvers were taken under the category of maneuvers, the presence of double parking and load/unload activity was checked under each time interval. Furthermore, the headway of massive bike flow t_{B1}, the critical gap t_c, service time for parking cars t_{B2}, and the total delay time T for inbound and outbound parking cars was measured respectively.

Survey Results

The characteristics of the selected eight test sections are shown in Table 1. Nearly 420 parking maneuvers were collected, and 8,566 e-bikes and 5,859 bicycles were observed. Both e-bike volumes and bicycle volumes were relatively high. From the table, it can be seen that the effective widths

Table 2. Statistical Description of The Critical Gap and the Service Time for Parking Cars

Category	Total influence time T(s)			Service time t_{B2}(s)			Critical gap t_c(s)	
	Mean	Minimum	Maximum	Mean	Minimum	Maximum	At the emerging point	At the diverging point
Inbound parking maneuver	23.9	8	63	22.9	7	63	3.7	12.5
Outbound parking maneuver	14.3	7	26	12.1	7	23	9.7	2.8

of the eight bicycle lanes vary from 2.20 to 4.50 m. The large variety of bicycle lane widths provides us with the chance to analyze the effect of the road characteristics on the capacity. The overall proportion of e-bikes is more than half (64%) and varies across sites from 54 to 68%. The results imply that e-bikes are an essential component of non-motorized vehicles and cannot be neglected. Some of the sites show relatively high parking demand (reflected by the parking maneuvers and the time influence rate). This means that the non-motorized vehicles observed in these sites may travel under the strong influence of curbside parking. Additionally, there were differences in the parking maneuvers and the time influence rate for the sites with similar effective widths. These differences may generate different influences on the capacities.

Calibrations Of Model Parameters

The model parameters consisting of $f(t)$, t_c, t_{B1}, t_{B2}, T, and the characteristics of the parking lot can be calibrated by the field data.

Characteristics of the Parking Lot and Non-Motorized Flow

Based on the data collected at the Emeiling Road, there are 11 parking stalls, the length and width of single parking spaces are 6.7 and 2.5 m, the type of parking space is parallel, and there were 50 events of inbound parking maneuvers and 41 events of outbound parking maneuvers in the survey period.

The free-flow-speed of non-motorized vehicles is 19.80 km/h and the t_{B1} is 1.8 s. The base capacity of a single bicycle lane calculated by Equation 1 is 2,000 bicycles/h. The average speed of parking vehicles (V_1) is 13.39 km/h.

Probability Density Function f(t)

Non-motorized vehicles usually travel as a massive group in one unit (*17*). The headway of non-motorized vehicle groups is a key factor that determines whether parking cars can pass through the conflict area. The field data of headways were collected at the upstream section, 25 m down from the parking lot. When the headway of non-motorized vehicles groups is larger than 0.4 s, there is little interaction among non-motorized vehicle groups. Therefore, 0.4 s is taken as a basis for delimiting non-motorized vehicles groups. A total of 485 samples were collected for calculating the probability density function $f(t)$:

$$f(t) = \frac{1}{0.806t\sqrt{2\pi}} e^{-0.005(\ln t - 14.732)^2}$$

Critical Gap t_c, Total Influence Time T and Service Time t_{B2}

The critical gap t_c, is defined as the minimum time interval in the non-motorized vehicles that allows the entrance of one parking vehicle. The gap that parking cars accept would be greater than or equal to the critical gap. Estimates of the critical gap were made on the basis of observation of the largest rejected and smallest accepted gap for the sections. The critical gaps for inbound and outbound parking maneuvers are listed in Table 2. The total influence time T and the service time t_{B2} for each event of an inbound and outbound parking maneuver were also observed. The statistical descriptions are given in Table 2.

Numerical Results

All model parameters are known. As there are a high number of non-motorized vehicles and parking maneuvers on the Emeiling Road, the paper takes this as the main case study for describing the computational steps of bicycle lane capacity.

Facts

- $t_{B1} = 1.8$s,
- $t_{B2}^{in} = 22.9$s,
- $t_{B2}^{out} = 12.1$s,
- $t_c^{in} = 12.5$s,
- $t_c^{out} = 9.7$s,
- $R_T = 35.48\%$,
- $W_e = 3.5$m,
- $eb_i = 0.67$,
- $V = 19.80$km/h,
- $V_1 = 13.39$km/h,
- $L_{n(i)} = 6.7\text{m}*i$,
- $M = 11$.

Table 3. Computations of the Average Conflict Probability and Adjustment Factor for the Time Influence Rate

Berth NO. (i)	Inbound parking m_i	Conflict probability P_C (%)	Outbound parking n_i	Conflict probability P_C (%)	Travel distance of inbound parking (m)	Actual travel time t_3^j (s)	Travel time for free flow $\frac{L_{n(j)}}{V}$ (s)	Travel distance of outbound parking (m)	Actual travel time t_3^j (s)	Travel time for free flow $\frac{L_{n(j)}}{V}$ (s)
1	17	24.59	16	59.02	6.7	1.80	1.22	65.6	17.63	11.92
2	11	36.29	10	59.02	12.4	3.33	2.25	59.7	16.05	10.85
3	3	44.14	2	59.02	18.2	4.90	3.32	53.8	14.46	9.78
4	3	50.15	2	59.02	24.2	6.51	4.40	47.9	12.88	8.71
5	3	55.02	2	59.02	30.2	8.12	5.49	42.0	11.29	7.64
6	3	59.05	2	56.68	36.1	9.71	6.57	36.1	9.71	6.57
7	1	62.52	0	52.7	42.0	11.29	7.64	30.2	8.12	5.49
8	3	64.89	2	47.9	47.9	12.88	8.71	24.2	6.51	4.40
9	2	64.89	2	41.97	53.8	14.46	9.78	18.2	4.90	3.32
10	2	64.89	1	34.23	59.7	16.05	10.85	12.4	3.33	2.25
11	2	64.89	2	22.69	65.6	17.63	11.92	6.7	1.80	1.22
$\bar{P}_C$		0.4780			Sum			281.72(s)		

Determining the Base Capacity

The average conflict probability caused by curbside parking $\bar{P}_C$ was calculated by Equation 6. The result was 0.4780 as shown in Table 3.

Therefore, the base capacity of the bicycle lane influenced by curbside parking can be calculated as

$$C_1 = \frac{3600 - Q_2 \cdot t_{B2}}{t_{B1}} = \frac{3600 - \sum_1^M m_i t_{B2}^{in} + n_i t_{B2}^{out}}{t_{B1}}$$
$$= 1562 \text{bicycles} //h$$

$$C_p = C_0 \cdot (1 - \bar{P}_c) + C_1 \cdot \bar{P}_c = 2000 * (1 - 0.48) + 1562 * 0.48$$
$$= 1791 \text{bicycles} / h$$

Determining the Adjustment Factors

The adjustment factor for bicycle lane width was computed using Equation 9.

$$f_w = \frac{W_e + 0.5}{b_e} = \frac{W_e + 0.5}{1.25} = 3.2$$

The adjustment for the time influence rate of curbside parking was computed using Equation 10.

The computational process is shown in Table 3.

$$f_T = 1 - \frac{\sum_{j=1}^N (t_3^j - \frac{L_{n(j)}}{V})}{3600} = 1 - \frac{281.72 * 3600 / 125 / 60}{3600} = 0.96$$

The electric bike adjustment factor was computed using Equation 12.

$$f_{eb} = \frac{1}{1 + \sum eb_i (n_{ec} - 1)} = \frac{1}{1 + 0.67 * (1.23 - 1)} = 0.87.$$

Determining the Actual Capacity Influenced by Curbside Parking

Therefore, the actual capacity of the bicycle lane influenced by curbside parking can be calculated as,

$$C = C_p \times f_w \times f_T \times f_{eb} \times f_j = 1791 * 3.2 * 0.96 * 0.87 * 0.55$$
$$= 2622 \text{bicycles} / h$$

Compared to the estimated results, the actual capacity for the same effective width of bicycle lane without curbside parking (W_e=3.5 m) is: $C = C_0 \times f_w \times f_{eb} \times f_j = 2000 * 3.2 * 0.87 * 0.55 = 3050 \text{bicycles} / h$, which indicates that the capacity would have a decrease of approximately 14.04% caused by parking manuevers.

If there was absolutely no curbside parking on the bicycle lane, the effective width would be 6.0 m (W_e = 6.0 m). Then the actual capacity could reach approximately 4,956 bicycles/h ($C = C_0 \times f_w \times f_{eb} \times f_j = 2000 * \frac{6.5}{1.25} * 0.87 * 0.55 = 4956 \text{bicycles} / h$), which indicates that the capacity would have a decrease of approximately 47% when influenced by curbside parking. The result implies that curbside parking occupies at least a 2.5 m width of the bicycle lane and decreases the capacity (47%). If there is not enough space for non-motorized vehicles, curbside parking should be prohibited to increase the bicycle capacity.

As there were a high number of parking maneuvers, the influence time rate, and e-bike volumes in Emeiling Road, the conclusion of a 14.04% expected reduction in capacity, which seems quite high, does not represent the average or general case. However, the case study is quite specific and only illustrates the influential factors of curbside parking on the bicycle lane capacity. Similarly, reductions in the capacity of the other sites were calculated: Zhongshan North Road

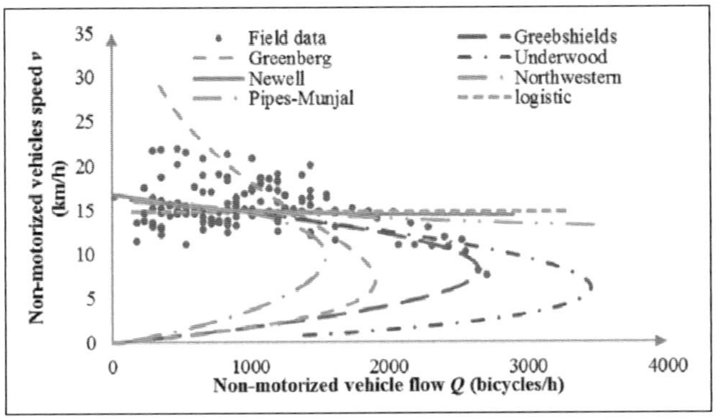

Figure 2. Relationship between non-motorized vehicles speed and flow.

A, 7.17%; Zhongshan North Road B, 6.68%; Zhongshan North Road C, 16.73%; Shuiximen Road, 18.79%; Guangzhou Road, 15.68%; Huaqiao Road, 16.03%; and Shanghai Road, 17.21%. The results show that there is a slight tendency that the reduction in capacity increases with the decrease of the effective widths (ranges from 4.5 m to 2.3). This can be explained by the decreasing space requirements, which can cause a longer duration of conflict between non-motorized vehicles and parking vehicles, decrease the travel speed of non-motorized vehicles through the encountering of parking maneuvers. Additionally, narrower effective lane width also increases the operation difficulty of parking cars and influences the time of parking maneuvers. According to the field data survey, the parking cars took 42 s to enter and 27 s to exit the parking stall on Shuiximen Road which had a width of 2.2 m. The average duration of parking maneuvers for narrow roads are twice those of a wide one.

Model Validation

Another main method of capacity estimation is to calibrate a relationship between three traffic flow variables: speed, volume, and density. Eight well-known speed-density relationship models were selected. The speed-volume relationship models were then obtained by using the fundamental formula $Q = kv$. The capacity is estimated by the maximum point of the speed-volume models as shown in Figure 2 and Table 4.

In which v and k are the speed and density of non-motorized vehicles in one minute intervals, m/s, bicycles/m^2, v_f is the free-flow speed, k_j is the jam density, v_m and k_m are the speed-at-capacity and density-at-capacity, v_b is the average travel speed under stop and go conditions, and n, m, θ_1 and θ_2 are parameters to be calibrated.

The bicycle lane capacity was calculated as the mean of the maximum traffic flows from eight models: $C = Mean(Q_{mi}) = 2735 bike/h$. The absolute error between the observed and estimated capacity is only 4.78%, which indicates that the proposed method is effective for calculating the capacity of the bicycle lane influenced by curbside parking.

Model Application

The proposed model can be used to evaluate the influence of different parking designs on the bicycle lane capacity and improve parking design. The frequency of parking maneuvers at the different locations of parking berths is taken as an example to present the model application for discussing the effect of curbside parking on the bicycle lane capacity. If the number of events of inbound and outbound parking maneuvers ranges from 11 to 50, the distributions of the capacity are shown in Figure 3. The inbound and outbound parking maneuvers indicate a negative effect on the capacity, so that the increase in the number of those bicyclists encountering parking vehicles could decrease the capacity. Furthermore, different positions of the parking berths have different influences on the capacity. As shown in Figure 3, the event of parking maneuvers at the parking berth No. 6 has the most influence on the capacity. By comparison, the influence of the parking berth No.1, 3, 9, and 11 is lower. This can be explained by the distance of the parking berth away from the access points, which can determine the continuous duration of parking maneuvers. Therefore, curbside parking along a bicycle lane, where allowed, should be set near the access points of physical separation, so that the parking car can enter or exit the parking stall conveniently. This may reduce the number of conflicts between non-motorized vehicles and parking vehicles. Fewer conflicts may result in less reduction of capacity.

Other influential variables can also be used to analyze the sensitivity of the capacity reduction in the same way. Most importantly, the capacity methodology based on the conflict technique can capture the effects of influential factors, which can modify the capacity. Therefore, the proposed model could be helpful in the operational analysis. For example, the estimated capacity could be a firm basis for evaluating the influence of curbside parking and improving the plan and design of curbside parking and bicycle lanes. Before application, the model should be estimated by using the specified field data. However, there are some limitations to the proposed methodology. First, it is a little complex for practitioners to utilize the methodology. Second, other influential variables such as psychological and cultural variables and bicycle parking maneuvers that could influence the bicycle capacity are not mentioned (even if they were discarded because of insufficient samples). Third, this study focused on the capacity of a bicycle lane with curbside parking but does not directly address the underlying safety issues (probably the most significant issues in the design of curbside parking and bicycle lanes).

Conclusions

A conflict technique based on additive conflict flow was applied to estimate the capacity of a bicycle lane under the influence of curbside parking. The actual capacity was established by adjusting a base capacity to reflect the influence:

Table 4. Single-Regime Speed–Density Models Proposed for Capacity Estimation

Name	Function	Result	Capacity
Greenshields (1935)	$v = v_f(1 - k / k_j)$	$v = 4.537 - 7.033k$	2634
Greenberg (1959)	$v = v_m \log(k_j / k)$	$v = 5.028\log(0.201 / k)$	1900
Underwood (1961)	$v = v_f \exp(-k / k_m)$	$v = 4.596\exp(-k / 0.568)$	3449
Newell (1961)	$v = v_f(1 - \exp[-\frac{\lambda}{v_f}(\frac{1}{k} - \frac{1}{k_j})])$	$v = 4.641[1 - \exp(-1.841 - \frac{0.026}{k})]$	2888
Northwestern (1967)	$v = v_f \exp(-\frac{1}{2}(k / k_m)^2)$	$v = 4.456\exp(-\frac{1}{2}(k / 0.159)^2)$	1542
Pipes-Munjal (1967)	$v = v_f(1 - (k / k_j)^n)$	$v = 5.530(1 - (k / 52.875)^{0.205})$	3548
Logistic (2011)	$v = v_b + \frac{v_f - v_b}{(1 + \exp[(k - k_m) / \theta_1])^{\theta_2}}$	$v = 4.092 + \frac{1.438}{(1 + \exp[(k - 0.005) / 0.158])^{7.279}}$	3282

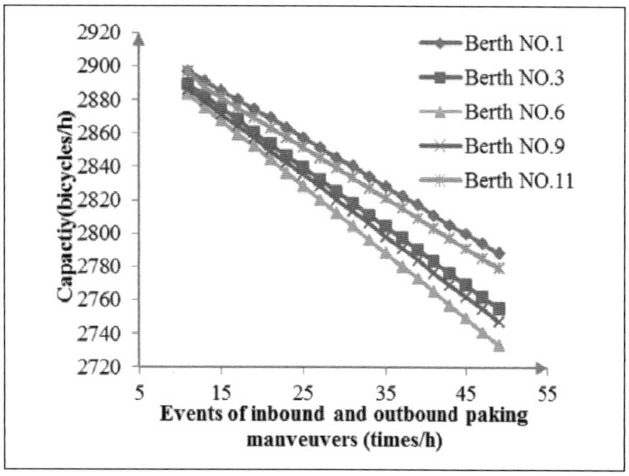

Figure 3. Capacity influenced by parking maneuvers at different berth locations.

effective lane width, the time influence of parking maneuvers, and the proportion of e-bikes. The paper used the field survey datasets from eight bicycle lanes in Nanjing, China for estimation and calibration. The parking maneuvers and the effects of curbside parking were also recorded. As the Emeiling Road had more parking maneuvers, this was taken as the main case study. The capacity of the bicycle lane was calculated, and the effectiveness of the proposed method was validated by the seven speed-density-volume relationship models.

Several important results were reported. Firstly, the average proportion of e-bikes is more than 50%, e-bikes as an essential component of non-motorized vehicles cannot be neglected. Secondly, the capacity of Emeiling Road with around a 65% share of e-bikes is 2,622 bicycles/h under the influence of curbside parking. The capacity of the bicycle lane would have a decrease of approximately 14.04% caused by curbside parking compared with the same effective width without curbside parking. Furthermore, the capacity would decrease by approximately 47.10% under the influence of curbside parking, compared with if there was absolutely no curbside parking on the bicycle lane. Thirdly, the proposed model based on the conflict technique is well-suited to the study of curbside parking. The absolute error between the observed capacity and estimated capacity is only 4.78%. Finally, the berths near the openings of isolation belt have less influence than those in the middle positions. The placement of parking berths are suggested to be adjacent to the opening of the isolation belt. Additionally, reduction of the bicycle lane capacity contributes to the decreasing space requirements and the conflicts with parking maneuvers. To reduce or eliminate the conflicts between parking cars and non-motorized vehicles, we recommended that the bicycle lanes should be placed on the outside of the curbside parking lot. It seems to be an easier solution to simply design the bicycle facility to have minimal effects from the curbside parking on the non-motorized vehicles. However, this type of design cannot eliminate the conflicts between parking cars and non-motorized vehicles completely, and may also generate other issues, such as the same conflicts between parking car and non-motorized vehicles, in addition to new conflicts between the drivers or passengers and non-motorized vehicles, such as open door problems, and so on. Therefore, it is necessary to study the different influences of inside and outside parking lots on bicycle lane capacities.

The proposed methodology has some limitations that may make the utility for practitioners marginal. More datasets are required in future work. Other adjustment factors should also be regarded, such as the types of curbside parking (angle or parallel), and the road gradient. It is also necessary to research the safety of bicyclists under the influence of curbside parking in the future. These studies will allow a systematic analysis of the characteristics of curbside parking. The findings and contributions of this paper are helpful and useful for practical countermeasures to improve the design and optimization of curbside parking on bicycle lanes.

Acknowledgments

This research is supported by the National Natural Science Foundation of China (NO.51408322, 51478111 and 51408145) and the Natural Science Foundation of Jiangsu Province (BK20150747).

References

1. Chick, C. *On-Street Parking: A Guide to Practice*. Landor Publishing, London, 1996.
2. *Highway Capacity Manual*. TRB, National Research Council, Washington, D.C., 2000.
3. Valleley, M. *Parking Perspectives: A Source Book for the Development of Parking Policy*. Landor Publishing, London, 1997.
4. Yousif, S., and S. Purnawan. On-Street Parking: Effects on Traffic Congestion. *Traffic Engineering and Control*, Vol. 40, No. 9, 1999, pp. 424–427.
5. Arnott, R., and J. Rowse. Modeling Parking. *Journal of Urban Economic*, Vol. 45, No. 1, 1999, pp. 97–124.
6. Marshall, W. E., N. W. Garrick, and G. Hansen. Reassessing On-Street Parking. *Transportation Research Record: Journal of the Transportation Research Board*, 2008. 2046: 45–52.
7. Portilla, A. I., B. A. Oreña, J. L. M. Berodia, and F. J. R. Díaz. Using M/ M/∞ Queueing Model in On-Street Parking Maneuvers. *Journal of Transportation Engineering*, Vol. 135, No. 8, 2009, pp. 527–535.
8. Box, P. C. Curb-Parking Problems: Overview. *Journal of Transportation Engineering*, Vol. 130, No. 1, 2004, pp. 1–5.
9. Ye, X., and J. Chen. Impact of Curbside Parking on Travel Time and Space Mean Speed of Nonmotorized Vehicles. *Transportation Research Record: Journal of the Transportation Research Board*, 2014. 2394: 1–9.
10. Homburger, W. S. *Capacity of Bus Routes, and of Pedestrian and Bicycle Facilities*. Institute of Transportation Studies, University of California, Berkeley, 1976.
11. Ferrara, T. C. *A Study of Two-Lane Intersections and Crossings under Combined Motor Vehicle and Bicycle Demands*. Civil Engineering Department, University of California, Davis, 1975.
12. Vagverk, S. *Bicycle Traffic Facilities (English translation)*. In *Swedish Capacity Manual*, National Swedish Road Administration, Stockholm, 1977, pp. 99–123.
13. Navin, F. P. D. Bicycle Traffic Flow Characteristics: Experimental Results and Comparisons. *Institute of Transportation Engineers Journal*, Vol. 25, No. 8, 1994, pp. 64–72.
14. Liu, X., L. D. Shen, and F. Ren. Operational Analysis of Bicycle Interchanges in Beijing, China. *Transportation Research Record: Journal of the Transportation Research Board*, 1993. 1396: 18–21.
15. Botma, H. Method to Determine Levels of Service for Bicycle Paths and Pedestrian-Bicycle Paths. *Transportation Research Record: Journal of the Transportation Research Board*, 1995. 1502: 38–44.
16. Stembord, H. Capacity Research and Applications: Country Report of the Netherlands. *Proc., 1991 International Symposium on Highway Capacity*, Rotterdam, Netherlands, 1991.
17. Wei, H., J. Huang, and J. Wang. *Models for Estimating Traffic Capacity on Urban Bicycle Lanes*. Transportation Research Board, Washington, D.C., 1997.
18. Ministry of Housing and Urban-Rural Development of China. *Code for Design of Urban Road Engineering*. CJJ37-2012, 2012.
19. Jin, S., X. Qu, D. Zhou, C. Xu, D. Ma, and D. Wang. Estimating Cycleway Capacity and Bicycle Equivalent Unit for Electric Bicycles. *Transportation Research Part A*, Vol. 77, No. 9, 2015, pp. 225–248.
20. Brilon, W., and N. Wu. Capacity at Unsignalized Intersections Derived by Conflict Technique. *Transportation Research Record: Journal of the Transportation Research Board*, 2001. 1776: 82–90.
21. Brilon, W., and T. Miltner. Capacity at Intersections without Traffic Signals. *Transportation Research Record: Journal of the Transportation Research Board*, 2005. 1920: 32–40.
22. Ye, X, J. Chen, and S. Gu. Conversion Coefficient of Electric-Bike into Bicycle on Urban Road Section. *Journal of Highway Transportation Research Development*, Vol. 29, No. 10, 2012, pp. 109–116.

The Standing Committee on Transportation in the Developing Countries (ABE90) peer-reviewed this paper (18-01143).